SOTHIS PRESS

Sothis Press aims to completely reimagine the public domain book with an emphasis on universal access to captivatingly designed print editions. We believe that technology will enable every single book ever published to be perpetually accessible in beautiful print form. Sothis is beauty in ubiquity.

www.sothis.press
1 Waterhouse Square, London, ECIN 2ST

The Distillation of Resins

Resinate Lakes and Pigments, Carbon Pigments and Pigments for Typewriting Machines, Manifolders, Etc

Victor Schweizer

SOTHIS PRESS

ISBN 978-1-0368-0908-9

The Distillation of Resins
is in the public domain.

This edition is published by
Sothis Press The company
is registered in England and Wales.
The Company Number is 14900389.

DISTILLATION OF RESINS.

RESINATE LAKES AND PIGMENTS.
CARBON PIGMENTS AND PIGMENTS FOR TYPEWRITING MACHINES, MANIFOLDERS, ETC.

A description of the proper methods of distilling resin-oils, the manufacture of resinates, resin-varnishes, resin-pigments, and enamel paints; the preparation of all kinds of carbon pigments and printers' ink, lithographic inks and chalks, and also inks for typewriters, manifolders, and rubber stamps.

BY

VICTOR SCHWEIZER.

With 68 Illustrations.

LONDON:

SCOTT, GREENWOOD & SON,

"THE OIL AND COLOUR TRADES JOURNAL" OFFICES,

8 BROADWAY, LUDGATE, E.C.

1907

This work is devoted to the description of the best methods of preparing resinous bodies, not only from our native pines, but from foreign trees and from fossil resins. As this preparation can only be properly effected by steam-distillation, the apparatus it requires and the methods and plant are fully described.

Besides the processes for obtaining the resins themselves, we describe the manufacture of the numerous products obtainable from them, including the resin lakes, resin lake pigments and resinates. These have never yet been properly appreciated, and special attention is given to them in this book.

As rosin oils are used to a greater and greater extent in the manufacture of printing inks, we were obviously compelled to go fully into the important matter of their manufacture. With regard to the preparation of carbon pigments, which, among other uses, are employed in large quantities for printers' ink, only brief notices scattered through technical literature have so far been available. We have, therefore, devoted a considerable part of our work to the exhaustive description of the manufacture of these important coloring matters with especial attention to suitable plant for the production of lampblack pigments.

Having regard to the constantly growing employment of type-writers, it appeared to us to be essential to append to the chapter on printing inks an account of the preparation of inks for type-writer ribbons, and for carbon manifolders. We call attention to the fact that the present work is the first publication dealing with this matter.

The present work gives a description of all processes concerning the obtaining and the preparation of rosin, according to the latest state of our knowledge, and the author entertains the hope that he has provided a book which will give every information that can be required by those interested in the trade.

CONTENTS.

Carbon Pigments.

Resins and their use for the Manufacture of Chemical Products.

I. INTRODUCTION.

The peculiar products of the vegetable kingdom which are generally known as resins are used in many different ways. Some of them are used for fumigating or for perfumery on account of their agreeable smell, as for example the benzoin resins. Others contain valuable colouring matters, such as lac, and the resin known as gamboge. A considerable number of resins are largely used just as they come from the plants in the preparation of lakes and varnishes, such as elemi, dammar, etc.

While all the resin belonging to one of the groups just named are used either alone or in combination with solvents, we are acquainted with a few other resins used for the preparation of chemical products. There is only one such resin of prime importance, by reason of the number of products made from it, *i.e.*, rosin, although amber and copal are used for the same purpose. Although by far the largest part of this resin comes from pine trees, some of it is produced by trees belonging to other botanical families.

So long as Europeans confined themselves to the product of their own trees, rosin exploitation was confined within somewhat narrow limits, as there were comparatively few trees. Conditions, however, became very different when the vast pine forests of the United States and of Canada began to put large quantities of very pure rosin on the European market.

It was then that greater attention began to be given to the preparation of chemical products from rosin, and the result was the establishment of a whole series of industries which have now reached a considerable development. As soon as it was discovered that rosin in combination with alkalies. gives products closely resembling the soaps made with alkali and fat, this resin began to be largely used in soap-making together with fat.

Another large industry was based on the fact that the dry distillation of colophony gives liquids which are excellent for lighting purposes, but the introduction of petroleum put an abrupt termination to this manufacture, as the rosin products were quite unable to compete with petroleum in the matter of price. In distillation to produce illuminating liquids, however, quantities of thick, oily liquids are also obtained which were at first almost waste products. The competition of petroleum naturally incited efforts to utilise these liquids, and they turned out to be the most valuable products of the distillation, as they are most useful in the preparation of lubricants and of printers' ink. They are now largely used with lampblack in the manufacture of that invaluable pigment, and the lampblack itself is made from these "rosin-oils."

Researches into the properties of the rosin soaps insoluble in water have shown that they serve well for the preparation of liquids which are often excellent substitutes for the expensive varnishes prepared from the fatty oils, and from which painters' colours of a good quality can be prepared. If we add to these uses the employment of rosin for making a capital illuminating gas and also brewers' pitch, we have said quite enough to indicate the great commercial importance of the chemical elaboration of rosin.

In the present work, all the uses so far discovered for rosin and for the products obtained from it are exhaustively described, and also the methods of a trade closely connected with the exploitation of rosin, namely the preparation of carbon pigments. Until now no connected description of this industry has been published.

In addition to rosin two other resins have been subjected to fractional distillation, namely the hard resins, amber and copal. The distillation is only fractional and is but preliminary to the preparation of hard varnishes. Little attention is paid to the other products of distillation, but as the running of copal and amber is closely analogous to the dry distillation of colophony, we have included it in this book. We are of opinion that this will be of service to the industry concerned, as the apparatus at present used in running is far from being free from faults of construction. In this division of our work, therefore, we shall give special attention to the clear explanation of the essentials of the running process, and indicate the proper arrangements for carrying it out.

II. ROSIN.

Among resins in general, rosin from pine trees takes one of the most important places, because it occurs in larger quantities than any other resin, so that it forms the raw material of several large industries. There is a whole family of trees, all of which yield rosin, but all of them do not belong to the genus Pinus, although that genus gives far more of the product to commerce than do trees belonging to the genera Larix and Abies. The following is an account of the chief trees from which we obtain rosin.

Pinus sylvestris, P. vulgaris, or P. nigra, the common pine (the Scotch fir) occurs in many varieties and is specially common in Alpine countries. This tree is the chief source of German rosin. The pines native to the limestone Alps are specially rich in rosin. They are known to botanists as Pinus nigricans.

French rosin also comes from P. nigricans, but most of it is obtained from P. maritima (the maritime pine) which is specially common on the south west coast of France, and of which large forests are there cultivated.

American rosin, now of the greatest importance to all rosin industries, comes from the swamp pine, P. palustris, which forms gigantic forests in the United States and in Canada. The quantity of American rosin put annually upon the markets of the world is so great that it has hit the European trade very hard and caused the failure of many factories. Thanks to the skilled manner in which it is obtained, the American is quite as good as any European product, and the vast output enables it to be sold at a price which has excluded the competition of European sorts in many places.

Besides the above mentioned trees, many other conifers yield resins, but we prefer to use the half-liquid balsams containing the resins as they are, without further preparation. These balsams are usually known as fine turpentines. The chief kinds are: Venice turpentine, from the trunk of a larch, Larix decidua: Hungarian turpentine from Pinus pumilio, obtained by cutting off the ends of the twigs and gathering the liquid which flows from the wounds; Carpathian turpentine, or Cedro-balsam, which is the product of Pinus cembra: Strassburg turpentine, obtained from the white pine,

Abies alba: Canadian turpentine, or Canada balsam, from the balsam pine, Abies balsamea: and lastly Chios turpentine, or Cyprus turpentine, from Pistacia terebinthina and P. vera of the Grecian Archipelago.

Ethereal oil and resin can be obtained from all these, but as already stated, the products are put on the market as they flow from the trees, and are used as they are, and not as raw material for the preparation of secondary products.

Collection of Rosin.

In the trees, the rosin often occurs in special vessels, known as the rosin-ducts, and the rosin in these appears to take no part in the life-processes of the tree. We occasionally also find larger receptacles in the tree, called rosin pouches. These do not usually contain rosin, but rather semi-fluid balsams consisting of essentia oil and rosin.

Most of the crude rosin is obtained by cutting the tree somewhat deeply and collecting the product as it flows out. In general the bark is removed from about a quarter of the circumference of the tree, and to a height of forty to sixty inches. A dish-shaped cavity is then cut at the foot of the trunk, and thin slips of wood, which are fixed slantingly in the stripped part of the tree, lead the outflow into the cavity, from which it is removed from time to time.

During the summer a yellow tough sticky mass with an aromatic smell oozes from the stripped part of the trunk. This mass is chiefly composed of a solution of rosin in essential oil. When the sap retreats in the late autumn the flow ceases till next spring, when a bark is removed higher up. In the course of years, the trunk becomes partly deprived of its bark up to a height of seventeen to twenty feet. Many trees will yield for twenty years. Finally, however, the summer outflow becomes so small that it no longer pays to exploit the tree any further.

The crude rosin, when poured into casks, divides into two layers, a small watery layer containing albuminous matter and mineral salts, and a far deeper lower stratum having a strong balsamic odour. This lower stratum consists of essential oil, rosin, and various other bodies which are to be regarded as accidental impurities. Splinters of bark and wood, the acicular leaves of the tree, as well as earth and stones, are usually found among these impurities. The watery layer is rejected as useless. To obtain pure rosin from the other product it must be subjected to distillation, so as to separate the rosin from the essential oil. The rosin residue is fused and filtered while still liquid, so as to get rid of the vegetable and mineral debris above mentioned. The lower stratum is brown and thick.

In former times the distillation was always executed with very primitive apparatus. With the better plant now in use a much better quality of oil of turpentine and a much paler and less decomposed rosin is obtained. The old apparatus consisted of copper stills heated over an open fire. Now when the water and turpentine are distilled off, the residual rosin assumes a thick and viscid consistency, and since it is a very bad conductor of heat, it is practically impossible, with an open fire, to prevent those parts of the rosin which are in contact with the bottom and sides of the still from undergoing decomposition. This not only darkens the rosin but causes the turpentine to be contaminated with rosin-oils resulting from the decomposition.

Distillation of Crude Rosin by Steam.

There is only one suitable process for distilling rosin, viz., with superheated steam. While water boils at 100 deg. C., the boiling point of turpentine is 160 deg. C. Hence steam at this temperature is able to drive off all the oil of turpentine. This temperature always gives a very pale yellow resin. But the turpentine oils partly volatilise in water-vapour at much lower temperatures. Hence if we make the crude rosin fluid by heating it to 100 deg. C., and then blow steam through it, the steam carries off all the turpentine. The rosin can then be dried in the still, and filtration alone is required to get it in a perfectly pure state. The use of superheated steam has the further advantage that it enables very large stills to be used. Many are in use which take ten tons of crude rosin at a time, and as four distillations can be done daily, forty tons a day can be handled with a single still of this capacity.

Fig. 1 shows the construction of a steam still for the distillation of crude rosin.

The cylinder C, made of thick iron plate, is so enclosed in a second iron cylinder M that the intermediate space is about four inches wide. When the steam-cock D is opened, steam enters this space, and rapidly heats up the contents of C. There is a second tap D_1, whereby steam can be admitted to the interior of C through a flat coil having a number of small holes on its underside. A second cock D_2 admits steam to the upper part of the interior of C, where the pipe has a plain opening. A large hole O is provided at the top of C for charging the still. It is filled with a steam-tight cover.

In the middle of the top of C is the hat-shaped dome H. Through this goes a rod governing the valve V attached to its lower end. A thread is cut in the rod which works in a nut which closes H, so that by turning the rod by the small wheel shown, the valve V can be either closed down, or lifted from its seat, so as to run the

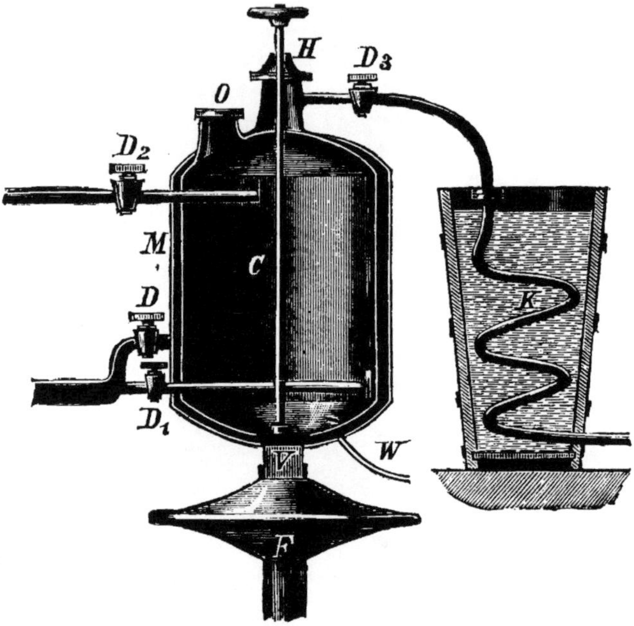

FIG. 1.

resin into the filter F. The cock D_3 regulates the tube leading from the still to the worm K in the condenser. The filter is screwed on to the bottom of the still, and consists of two conical iron vessels put base to base, so as to hold a wide-meshed piece of wire gauze over which a linen cloth is stretched.

When C has been about three parts filled with the crude rosin steam is admitted first to the jacket. The rosin in C rapidly becomes liquid. The condensed water is drained from the jacket by the pipe W. When the rosin is all melted, D_1 is opened, which sends the steam into C where it streams through the resin from the holes in the coil. This rapidly causes a distillate to appear in the receiver,* and this consists of oil of turpentine and water. This part of the process is continued until nothing comes over but water, whereupon D_1 is shut. The operator then waits until water ceases

* Not shown in the figure. It receives the condensed products from K.

to flow from the worm. This shows that the rosin in C is dry. He then shuts D_3 and lifts the valve V by the wheel above H, and at the same time the cock D_2 is very gradually opened. The pressure of the steam thus admitted drives the fused rosin through the filter. It is of very great importance to admit the steam above the rosin gradually, for too sudden a pressure would probably tear the filter cloth. The pressure is regulated with D_2 in such a way as to get a uniform flow through the filter into the receptacle below it. When nearly all the rosin has run out, D_2 and V are shut, and the still s at once filled with a fresh charge of crude colophony. The second and subsequent distillations take less time than the first, as the apparatus is already hot when the crude material is introduced.

When the temperature of C is so high that distillate begins to flow from the worm, it has a milky appearance, being an emulsion of oil of turpentine in water. On standing in the receiver this milky liquid separates into two sharply marked layers, water below and turpentine above. The water is then drained off through a cock in the bottom of the receiver.

This method of separating the turpentine and water cannot however be recommended. Oil of turpentine absorbs oxygen from the air with great avidity. By so doing it thickens, becomes yellow

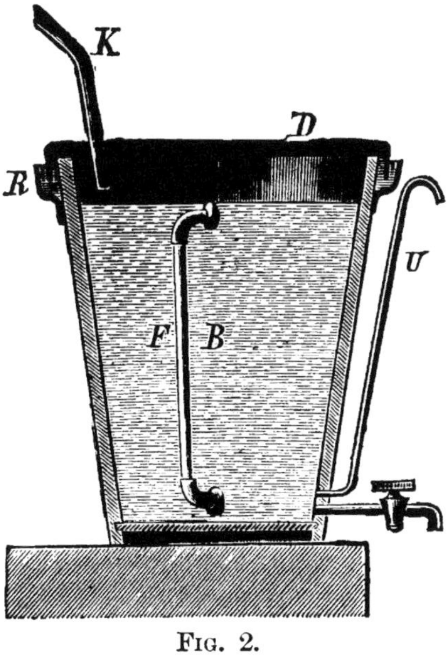

FIG. 2.

and acquires great bleaching power, as it contains the absorbed oxygen, at first in the form of ozone. If an organic dye is dissolved in oil of turpentine which has been left for some time in contact with the atmosphere, it will be found that the dye is nearly entirely destroyed by the ozone in the oil. There is a very simple apparatus shown in Fig. 2, whereby this injurious action of the air can be entirely prevented.

B is the vessel in which the distillate is received. Its upper edge is provided with an iron gutter R, running all round it. The gutter is about one and a half inches wide and four inches deep. This gutter receives the edge of the cover D. The end K of the worm passes through a hole in D. A gauge glass is fixed to the side of B, so that the amount of liquid in B can be at once perceived. A tube U comes from the bottom of B, and opens freely as shown, with a downward curve. At the beginning of the distillation R is filled with clean water so as to make the cover air-tight. But K must not fit air-tight into the hole in D, for the air must be allowed to escape from B, as it gets filled with distillate. After a time the depth of the distillate is such that water begins to flow from U and then flows away continually as fast as it distils over. When there is little else in B but oil of turpentine, the contents are allowed to stand for a few days to enable the separation to become perfect, and the small amount of water remaining is run out by a cock at the bottom of B.

The first accumulation of turpentine on the surface of the water partly evaporates, and its vapour affords a complete protection to the liquid portion from any traces of oxygen present. The oil of turpentine is thus obtained perfectly colourless, free from absorbed oxygen, and having strong refractive power and an aromatic smell.

The rosin which has passed the filter forms on cooling a pale yellow mass, occasionally showing a reddish tinge. It is usually as transparent as glass, and has neither smell nor taste. It is however very brittle, and soon loses its transparency with the shocks it experiences in transport, as it becomes traversed by numberless small cracks.

Pure rosin dissolves freely in essential oils and in caustic soda or potash lye. When it is boiled with a solution of carbonate of soda or potash, its resinic acid expels their carbonic acid, and forms a soda or potash rosin soap. Hence its great use in soap-making. It is also used for the manufacture of rosin oil for making varnishes and resinate colours; and when burnt in a limited supply of air, it gives a lampblack which is remarkable for its fine sub-division and for its lustrous black colour, in both of which respects it excels many other forms of the pigment. Besides yielding this black pigment rosin gives an illuminating gas of much greater candle-power than coal-gas.

The residue in the filter of course retains a little of the rosin. It can, however, be used to great advantage as a fuel. As already stated, it is virtually impossible to prevent decomposition of the rosin in the still when the latter is heated over a naked fire. In this way we get, often in considerable quantities, a black easily fusible mass known as pitch. The only use of it is in caulking wooden ships. It was formerly used by shoemakers for waxing their thread.

Brewers' Pitch.

Much rosin is used for the preparation of this substance. It is used in breweries to produce a smooth lining inside casks whereby they are made much easier to clean. Brewers' pitch consists of rosin mixed with substances intended to have an effect upon the taste and aroma of the beer.

The simplest way of making a good brewers' pitch is to fuse pure American rosin with careful stirring to prevent the rosin from burning to the bottom of the pan. During the stirring sufficient rectified oil of turpentine is added gradually to make the mass free from brittleness when cold. In using the pitch a new cask is heated with a blast of hot air, and when thoroughly heated a proper quantity of melted pitch is poured in and spread over the sides of the cask by rolling the cask about until it is quite cold. To repitch a cask, set fire to the old pitch, let it burn some time, extinguishing it by putting the head on the cask, and repitch as a new cask.

Many makers of brewers' pitch use a certain amount of rosin oil instead of rectified oil of turpentine to make the mass supple. The rosin oil used must be rectified, as the crude oil would impart its penetrating smell to the beer. Others give the necessary suppleness to the rosin by fusing it with a soap made by boiling part of the rosin with caustic soda lye. Many recipes for making brewers' pitch advise the addition of soot or ochre, or of beeswax. All such additions are not only absolutely superfluous, but are very injurious to the beer kept in the casks. This is especially true of soot, which contains not only carbon but very considerable quantities of empyreumatic bodies of characteristic taste and smell, eminently calculated to spoil the flavour of beer.

III. HARD RESINS.

Most of the resins largely used in industry are wanting in hardness, and can be easily scratched with the finger nail. Such resins are thus commonly called soft resins, a general term including a large number of resins. Ordinary rosin is one of the soft resins, but it is of course harder than those which have only a semi-solid character at ordinary temperatures.

There are a few resins much harder than rosin, and they are therefore called the hard resins. Two of them are never found except in the fossil form, *i.e.*, as masses buried in the earth. They are true resins, and of vegetable origin, although the plants that formed them are now extinct. They include amber and certain kinds of the large and important group known as the copals. Both amber and copal are of the greatest commercial importance, as from them are made varnishes which are superior to all others in durability and resistance to atmospheric and chemical influences.

Amber.

This fossil resin is called Bernstein in Germany, from its inflammability. Its powers of attracting light bodies when rubbed excited great interest among the civilised nations of antiquity, and our word electricity is derived from the Greek word for amber, elektron. Aristotle had a perfectly correct idea of the origin of amber, which he declared to be the resin of a tree, but after his time many fables were circulated on the subject. Some declared it of animal origin, others said that it was condensed "sun-ether," others that it was a wax produced by a kind of ant. No scientific proof of the vegetable origin of amber was forthcoming until the end of the eighteenth century.

The origin of the great variety of opinion as to the origin of amber, as well as of the singularity of the theories, is to be sought in the fact that it was known to the ancient nations of Western Asia and was an important article of trade with them. When we remember that all the amber these people had came from the countries on the Baltic, we see easily that many false legends were likely to accumulate around a substance brought such great distances, as it was to the interest of every dealer in it to keep its origin a secret. Even the Romans who carried their arms into the heart of Germany had no knowledge of the origin of amber, which they called succinum.

When once the vegetable origin of amber had been clearly established, greater interest was taken in the deposits of it and researches in them added largely to our knowledge of fossil insects, as enormous numbers of them have been found perfectly preserved in the resin.

As regards the plant or plants which produced the amber, they were undoubtedly coniferous, but it is yet undecided whether the genus was Pinus or Picea. Provisionally, the name Pinites succinifer has been given to the tree producing amber. It is supposed that that part of the earth's surface now covered by the Baltic was a mountainous country rich in limestone at a time when great geological changes made an end to the Cretaceous period. The climate of this mountainous country was tropical as is shown by the plants found fossil in the district. The forests covering the mountains consisted principally of conifers among which there were large numbers of the amber-tree.

As in the course of years resin was formed in large quantities in these primeval forests and, falling to the ground, remained unaltered after the trees had perished and been replaced by younger ones, it is easy to see that in time immense quantities of amber must have accumulated. When the country became covered by the sea, the whole of the vegetation was destroyed. The trunks of the trees were swept away by the waves and the amber remained buried in the sea-sediment. This in time became a hard rock, known from its colour as blue earth, and it is to be regarded as the original amber, bearing stratum. In later geological changes, occurring in the brown coal period, the blue earth was displaced to a large extent. and the so-called "striped sand" was laid down. In this, too, we find amber embedded. Amber is also found in the Sarmatian and North German plains, and as here it is a long way from the sea, we infer that it was brought thither from the north by the ice of the glacial period, and left behind on the melting of the ice.

Storms cast up amber on the coasts of the Baltic. It is also sometimes fished up from the bottom of the sea. At present, however, by far the largest part of the amber of commerce is obtained by mining in the alluvial deposits on the shores of the Baltic and washing the sand and mud away from the amber.

Crude amber has usually a reddish brown crust covered with a yellow dust. It is considered that this crust is a result of decomposition of the amber proper, produced by the action of the water. It is removed from the larger pieces by scraping before they are put upon the market. Commercial amber appears either as a mass with a characteristic yellow colour, or as a mass partly transparent and partly streaked and clouded with turbid portions. The transparent part is the compact amber, while the other part owes its appearance to a multitude of tiny air bubbles entangled in the resin.

A large number of sorts and qualities of amber are recognised in the trade. The larger the pieces are, and the more beautifully they are marked, the higher the price, but when it is merely a question of working up the resin into varnish, these considerations naturally fall to the ground, and for such purposes we use the smaller pieces and also the shavings (*rasura succini*) obtained when the large pieces are turned on the lathe to make pipe-mouthpieces and ornamental objects.

Amber is the hardest of all resins, but its specific gravity varies somewhat widely. Pieces may be found on the one hand lighter than water, having a specific gravity of 0·98 only, while others may be as high as 1·2. Hence some specimens float in the sea, and some sink.

As regards chemical composition, amber must be regarded as a mixture of at least three different resins, and various but always small quantities of bituminous bodies are also present. Another invariable constituent is succinic acid. The three resins can be fairly and easily separated by their different solubilities in alcohol, ether, and chloroform. As a whole, amber is insoluble in every known solvent. Amber will endure heating to far higher temperatures than any other resin, but even with every precaution it cannot be run without undergoing a partial decomposition in the form of a dry distillation, giving off amber oil (*oleum succini*), and leaving behind a residue known as amber colophony, which will dissolve in various solvents, and serves for the preparation of amber varnishes. These varnishes cannot be made from the undecomposed amber as that is insoluble. There is one important difference between the heating of rosin and heating amber. The rosin oils given off by rosin are of great commercial value, and are prized for their own sakes. The amber oil is of no use by itself, and is only worth collecting to add to the varnish made from the residue of the distillation. The residue of the rosin distillation, pitch, is of very little use; so that in the one case the material is heated for the sake of the volatile constituents of the resin, and in the other for the sake of the non-volatile portion of it.

Copal.

The second of the hard resins, copal, is one of great importance to the varnish trade. Of amber there is only one sort, and the only differences lie in the size and beauty of the pieces, but there is a considerable number of kinds of copal. The reason of this is that there are many very different plants which yield the resin, so that it is produced in varying degrees of hardness and solubility and differing in other properties as well. The primary distinction made in commerce is that between hard and soft copals. The soft kinds are very often called gum animi, but English merchants apply the term indiscriminately to every sort of copal, hard or soft.*

No absolute line of demarcation can be drawn between hard and soft copals. The hardness of both lies between that of calc spar, and that of gypsum, they scratch the latter and are scratched by the former, except that some of the South American copals are scratched by gypsum. They are the softest kinds we know. The specific gravity of the copals also varies greatly. The lightest are hardly heavier than water, having a specific gravity of 1·045. In the heaviest the specific gravity may be as high as 1·139.

Although the largest amount of copal is yielded by still living trees, quantities of it, including the best sorts, are found as fossils in the earth, and it is probably that the trees producing these kinds belong to an earlier geological period and are now extinct. This fossil copal which occurs in pieces from the size of a pea to the size of the hand, is found in West Africa (Sierra Leone, Benguela, and Angola), in a special stratum consisting of sand, loam, and marl, and occurs at an average depth of ten feet. It is not improbable that this stratum is not the original burying place of the copal, but that it has been carried to its present situation by floods and buried in the sediment from the water. If this opinion is correct, the occurrence of copal is analagous to that of amber, and the copal is undeniably fossil.

The following are the most important kinds of copal in commerce:

East African copal, Zanzibar and Mozambique sorts.
West African copal, Angola, Benguela, and Sierra Leone sorts.
Kauri copal, New Zealand and New Caledonia sorts.
Manila copal or Sandaron.
South American or Hymenæa copal.

---

* We question this. We believe that, on the other hand, animi and copal are considered in England to be two different resins.—Tr.

The most valuable of the copals are the East African sorts. They are the hardest of all, and have the highest melting point. Nothing certain is known as to the trees of origin. Two which are said to produce this East African copal are Trachylobium mosambicense and Trachylobium Hornemannianum. The former tree is confined to East Africa, but the latter is native in South America also.

As above stated, West African copal is found fossil only, and nothing is yet known of its origin. Kauri or Cowrie copal is produced by Dammara australis and Dammara ovata, and this sort forms at present the largest part of the copal of commerce. These trees are so rich in resin that lumps weighing one cwt. have been found at the foot of old dead trunks. New Zealand copal is usually not uniformly coloured, but is traversed with light and dark streaks, has an aromatic smell and clings to the teeth when chewed, while the other copals are as gritty as sand when masticated.

Manilla copal comes from Vateria indica, a tree which grows in the Philippines and throughout tropical Asia.

South American copal is chiefly produced by two trees, Hymenæa coubaril and Trachylobium martianum. It is obtained like resin by making incisions in the bark and collecting the resin as it exudes from the wounds. A good deal of the copal is found on fragments of bark and especially on roots a short distance below the surface. The roots are often covered with a thick crust of the copal.

The African copals are the chief hard copals, but they show a good deal of variety, and much practice is necessary to enable one to distinguish the various kinds with certainty by the eye.

East Indian copal is either colourless or yellow, or dark red, and usually has a warty exterior. The West African copals dug out of the ground are sometimes crusted over with loam and sand. Hard copals can be polished, and the ground surfaces resemble those of amber similarly treated.

The soft copals come from South America (West Indian copal). They occur in pieces from the size of a pea to that of the fist, and having a spherical or tear-like shape. They may be milky white or pale yellow and transparent. Their hardness is so small that even hard rubbing with a coarse woollen cloth will wear down the copal perceptibly. New Zealand copal behaves in a similar way, as do many kinds of Manilla and Borneo copal.

The great variety in the copals is well seen in their behaviour to heat. While the soft sorts begin to melt as low as 180 deg. C., the hardest kinds require as much as 240 deg., a higher temperature than is required to run amber.

Soft copals can be dissolved in chloroform or in absolute alcohol, but to get a clear solution it is essential to powder the copal finely and to soak it in ether till it has completely swollen up. As, however, both chloroform and absolute alcohol are expensive solvents, they are rarely used in the preparation of copal varnishes, and copal will easily dissolve in the ordinary varnish solvents when it has been run or melted, *i.e.*, subjected to a partial dry distillation which drives off certain volatile bodies.

The great differences in the temperatures required by the different kinds of copal, make it essential to ascertain on a small scale the heat required by a new sort of copal, before treating it in large quantities.

IV. DISTILLATION OF THE HARD RESINS.

In distilling amber and copals the object in view is very different than in the case of rosin. In the latter case the object is to get as much distillate as possible (rosin oil) for the residual pitch is very little value. In the former case the object is to volatilise as little of the resin as possible, as although the heating is necessary to make the resin soluble, the resulting distillate is of small value,

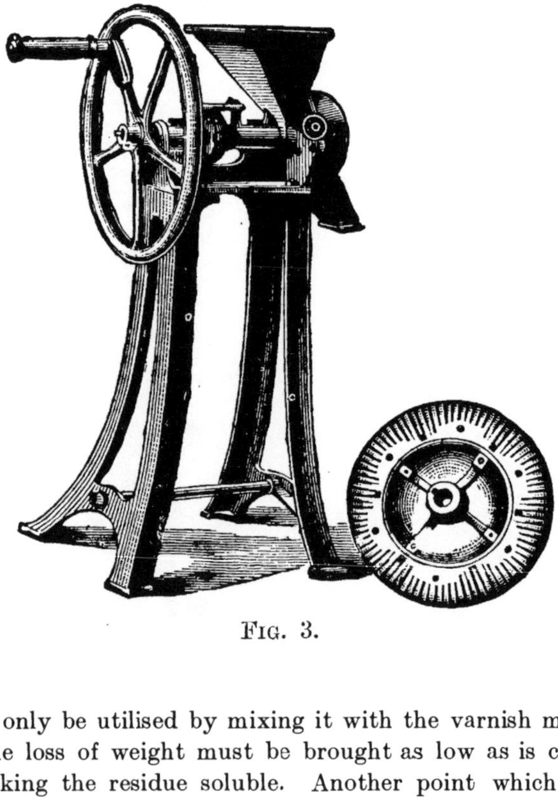

FIG. 3.

and can only be utilised by mixing it with the varnish made. In short, the loss of weight must be brought as low as is consistent with making the residue soluble. Another point which has also to be attended to is to get that residue of as light a colour as possible, for then only will it give a pale varnish and pale varnishes are much more prized than dark ones. As copal naturally occurs in all colours from pale yellow to a dark reddish brown, it is advisable to sort the resin into three kinds according to colour and to work up the three lots separately. As, too, both amber and copal are bad conductors of heat, it is of importance to distil together pieces of about the same size. To get such pieces the copal must be broken up by a suitable machine, such as is represented in Fig. 3.

This mill acts by means of two rotating discs having triangular teeth on their circumference and the distance between which is

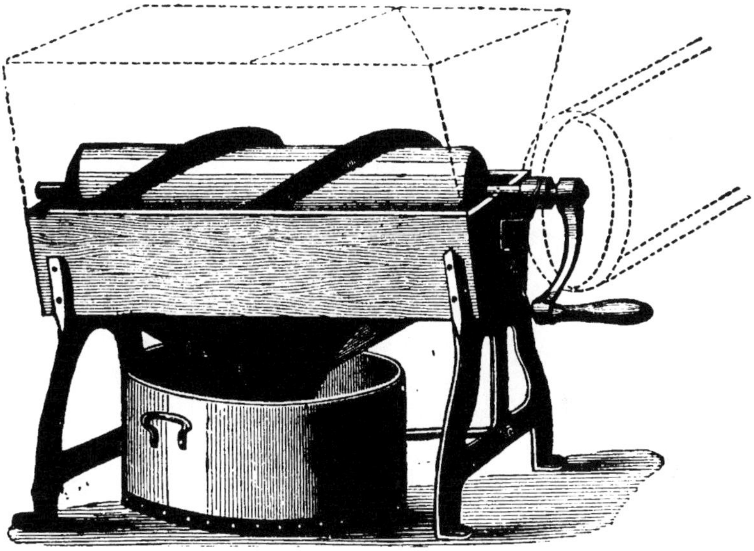

FIG. 4.

adjusted according to the size the pieces of resin required. The mill delivers small lumps of uniform size, but on account of the brittleness of the resin there are also pieces smaller than intended and a good deal of dust. The output of the mill must, therefore, be sifted by the machine shown in Fig. 4.

The brushes attached to the roller brush the dust off the pieces to which it adheres rather tightly. The smaller pieces and the dust which passes through the sieves are worked up separately.

Fig. 5 represents a copal mill which gives pieces of uniform size with very little dust. It can be adjusted so as to give pieces as

FIG. 5.

small as a lentil or as large as a big bean, and the small amount of powder is easily separated by sifting.

DRYING COPAL.

Certain copals have the property of becoming soluble after an exposure for three or four days to a high temperature. As, however, this temperature is never below 180° C., the word "drying" seems unsuitable. "Roasting" would be better. The process can be carried out both with small pieces and with powder but it must not be done in metal vessels, which greatly darken the copal. The best vessels seem to be large and very shallow pans of stoneware or porcelain, or well-enamelled iron in which there is no chance of the resin touching the metal. The apparatus required is one in which a temperature of 180 to 200° C., can be kept up for a long time. It should be possible to bring the temperature up to 300, as then the apparatus will serve for the treatment of amber.

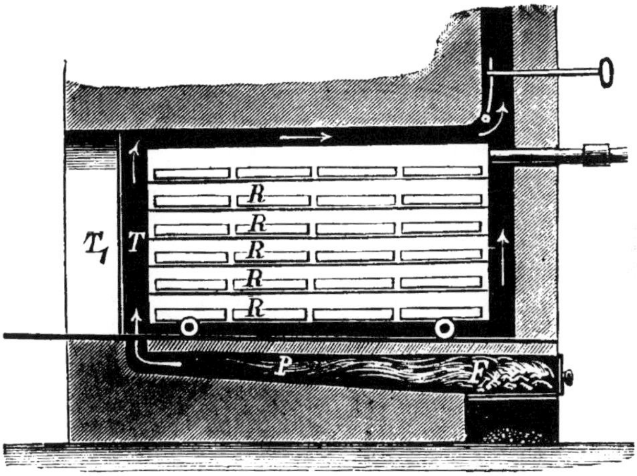

FIG. 6.

The apparatus shown in Fig. 6 is an excellent one for these purposes. Its construction greatly resembles that of a muffle furnace such as is used for enamelling. It consists of a masonry or brickwork chamber with a thick earthenware floor P below which is the fire F stoked with coke. At the end of the chamber furthest from the grate there is a slot in the floor, through which the furnace gases enter the chamber and pass in the direction indicated by the arrows through the chamber and out at the slot S. A damper R put into this slot near the outet of the combustion gases insures accurate regulation of the draught in the fire, and hence of the temperature attained in the chamber The chamber contains a box of thick iron plate, nearly filling the space so that the sides of the box are only a few inches from the walls, and the box is therefore constantly enveloped by the hot gases from the fire. This box, which is provided with an airtight door at T, rests on four small wheels running on rails so that it can easily be withdrawn from the chamber. A second door T_1 closes the other end of the chamber and is luted with clay inside. The tube leading from the box can be screwed on to a condensing worm so as to collect the volatile products. The inside of the box is fitted with strong wire netting shelves on which the vessels containing the resin are placed. To make the greatest possible use of the space these vessels are quadrangular and shallow, and the copal is put in them to a depth of one or two inches. The depth must not be too great as the dust of the copal cakes the mass whereby the drying is delayed, but the depth may be increased if the copal-powder is mixed with its own weight or thereabouts of clean white sand, which has no action on the copal and makes it porous. When the copal is dissolved the sand remains to be filtered off and can be used again for the same purpose.

The Electric Thermometer.

In order to regulate the temperature of the box an electrical thermometer is fixed in the middle of it, and this rings a bell as soon as the heat has reached a certain point. When the bell rings the

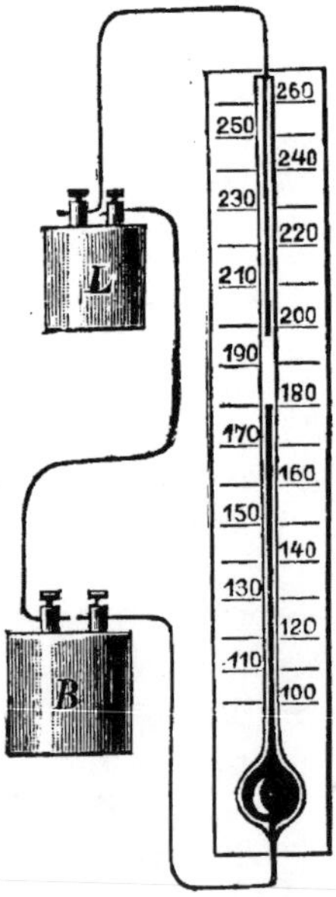

Fig. 7.

damper must be partly closed so as to prevent a further rise of temperature. A very simple but effectual thermometer of this kind is illustrated in Fig. 7.

This consists of an ordinary mercury thermometer with an open upper end through which a platinum wire nearly fitting the bore is inserted into the tube. A platinum wire is also fused into the bulb so as to come into contact with the mercury. When the thermometer is to be used the upper wire is adjusted in the tube so that its lower end stands opposite the mark on the scale showing the temperature which is not to be exceeded, say 200° C. One wire is connected with the negative pole of a battery, the other is its positive pole. As long as the temperature determined is not reached, the mercury leaves the circuit open, but as soon as it touches the upper wire the circuit is made and the bell rings and continues to ring until the temperature has so far fallen that the mercury has left between itself and the upper wire a space sufficient to interrupt the current.

As soon as the box is filled with dishes full of copal, its doors are luted and it is run into the chamber by means of the rails. The chamber having been closed, the fire is lighted and a good fire may be made at once as it takes some time to heat up the chamber and the box. The fire is kept up until the electric bell rings whereupon the damper is nearly shut and the fire is kept as high as possible without starting the bell again. The tube D having been connected with a condensing worm a small quantity of distillate is collected which never exceeds from 2 to 2½ per cent. of the weight of the copal treated. It is almost entirely composed of pure water and so long as that is the case it is evident that the heat has not been excessive, for had it been so the distillate would have contained oils resulting from the decomposition of the resin.

No change is perceptible in the roasted copal, but it has become far more easily soluble. As the distillate in a properly conducted operation is practically nothing but water, we might ascribe the previous insolubility of the copal to its containing water. Nevertheless, experiments have shown that the roasting may have other effects on the copal than a merely drying action. On drying copal powder over sulphuric acid *in vacuo* we find that after a few weeks the copal has lost several per cent. in weight, and that the sulphuric acid has gained in weight exactly as much as the copal has lost. Hence the loss is attributable to the evaporation of water from the copal and the gain to its absorption by the acid. In this experiment every trace of water is abstracted from the resin. But nevertheless this drying of the copal at the ordinary temperature does not make its solubility any greater than it was before. This experiment shows therefore that the lengthy roasting causes a change in position of the smallest parts of the copal with the result that there is an increase in the solubility.

The apparatus used for roasting the copal can also be employed for the distillation of the resin, by heating so strongly that after the coming over of the water a distillate of a yellowish oil floats upon the water which preceded it. These drops consist of volatile products of distillation produced by the decomposition of the copal by heat. The fire is in this case regulated in such a way that the oil distils over at a uniform rate and the distillation is kept up until as much oil has been got as is wanted. To prevent the iron of the box and the shelves in it from being attacked by the vapour of the copal oil, whereby the oil itself would be made of a dark colour, it is a good plan to varnish the whole of the interior with the best thick amber varnish before the box is first put in use. This protects the iron permanently from the vapours.

As rather a high temperature is required for the distillation of copal and a still higher one, *i.e.*, up to 400° C., for amber, the iron box must be carefully made. It should be constructed of strong boiler plate and rivetted hot.

The Distillation of The Resin.

If we subject copal or amber to distillation to make it easily soluble, we must naturally take care that the process is not carried further than is necessary to secure the object in view. Over distillation will result in a loss of resin, for which the oil obtained will not by any means compensate.

Although it cannot be denied that the various copals behave very differently on distillation, and that more oil must be distilled from a hard than from a soft copal to obtain sufficient solubility, yet there is a limit which may be taken as common to all the sorts of copal. Most manufacturers consider that the loss of weight undergone by the copal should not exceed 25 per cent. As it is impossible to keep stopping the distillation to ascertain the loss the copal has undergone, some other means must be found of gauging the percentage lost. This is done by estimating the amount of oil in the distillate, and from this the condition of the copal in the still is inferred.

Violette has made careful determinations of the changes in the solubility of copal brought about by distillation for various periods of time, and it was from his experiments that we were enabled to fix the loss of weight to be undergone by the copal The following figures give Violette's results for 100 lb of copal :—

Loss in weight lbs.	Oil obtained lbs.	Behaviour of the residue due to oil of turpentine.
3	3	Insoluble.
9	8·5	,,
10·5	10.2	,,
16	15·7	,,
20	19	Slightly soluble.
22	21·3	More soluble.
25	24·5	Very soluble.
28	27·1	,,
30	29	
32	31	,,

An examination of these figures shows that the yield of oil obtained is somewhat less than the loss of weight in the copal. The reason of this is that some of the products of the distillation consist of gases which escape into the atmosphere. The further the distillation is carried the greater the proportion of gaseous products. With a total loss of 25 per cent. (just the proper limit to make copal soluble), the loss in gas is only $\frac{1}{2}$ per cent., while with a loss of 28 per cent. it is $\frac{9}{10}$ per cent., and with one of 30 per cent. 1 per cent. We may now consider that we have made our copal sufficiently soluble when the weight of oil distilled over is one-quarter of that of the original resin. Beyond this limit it is useless to proceed, for thereby we duly diminish the yield of soluble product without securing any compensating advantage. To hit the point properly the same weight of copal or amber should always be taken for distillation, and the receiver should be of a corresponding size. This receiver must be graduated so that the amount of distillate in it will show at once how much of the resin has been decomposed. The best receiver is a tall glass cylinder graduated in the following way, assuming that 100 kilograms of copal are distilled at a time :—

Put exactly 800 grammes of water into the cylinder and mark the level of the water on the outside with a glazier's diamond Then put in 800 grammes more and mark the new level in the same. way, and proceed in the same way until 20 kilograms of water have been put into the receiver. Each division of the scale on the glass will then correspond with sufficient accuracy to 1 per cent. of the copal. The volatile oil produced by the distillation of the copal has a specific gravity of about 0·8, so that 1 per cent. of the 100 kilograms of copal, viz., one kilogram, corresponds to 800 grammes of water. For amber also, supposing a distillation of 100 kilograms at each batch, we proceed exactly in the same manner, but making the marks of the cylinder for 900 grammes of water instead of 800 as the specific gravity of the volatile oil in amber is 0·9.

Distillation of Hard Resins with Superheated Steam.

The distillation of the hard resins can also be done by means of superheated steam, which has the advantage that all danger of overheating the resin and so getting a dark residue can be avoided and that as soon as the result is achieved the resin can be at once removed from any further exposure to heat. An apparatus in which resins can very conveniently be distilled by superheated steam is shown in Fig. 8.

Fig. 8.

The still C consists of an oval cylinder made of strong copper-plate. The inside is painted over before the still is used for the first time with the best amber varnish. The still has two outlet pipes, D and K. The former carries the volatile products to the condenser while the latter, in which there is a wire sieve, is used for drawing off the melted resin. The other end of the cylinder has a well-fitting lid which is luted on during work. The cylinder C is enclosed within another C_1, so that the free space between them is a few inches across. C_1 is of iron, and is cleaded with asbestos to prevent loss of heat. Superheated steam enters this jacket through the tube W and leaves it by W_1, which also serves for the escape of condensed water. The apparatus is placed rather on a slant so that the fused resin will easily escape through K.

When the apparatus is ready for use the steam is sent in at W until the first drops of distillate appear, and fused resin begins to come from K. We then know that the contents of the still have been brought to the temperature at which the decomposition begins, and we regulate the flow of steam so that the distillation proceeds at a uniform rate. When the process is over the steam is shut off, and the whole apparatus is allowed to get quite cool before being opened. We can have the still large enough to take from 10 to 15 kilogrammes of resin for a charge. The distillation of this quantity is soon finished. For working on a larger scale, it is best not to have a larger still, but several of the smaller size, so that work need not be stopped for the still to cool. If six stills are started one after another the first is generally cool enough to be recharged by the time the sixth is started and can be recharged and restarted. In this way the distillation can proceed uninterruptedly, and only one still being idle at a time, a considerable saving in steam is effected, a point of much importance.

An apparatus which gives good results with an open fire, especially with amber, is shown in Fig. 9.

FIG. 9.

The boiler is set slanting in a fireplace. In the boiler is the rousergh the axle of which goes through a stuffing box at the top of the boiler. The top also carries a pressure gauge *f*, a safety valve *e*, and an outlet pipe *m* for the vapours. At the lowest part of the boiler at *k* are the sieve and overflow pipe for the melted resin; *b* is the manhole for charging in the amber.

Having fired up we send steam at two atmospheres pressure through the tube *d* to drive all the air out of the boiler. The stirrer must be kept going throughout the operation, and the volatile oils escape by *m*, while the used amber is flowing out at *k*. The stirrer is not of much service. As long as the amber is well below the temperature at which it begins to decompose, the pieces are kept moving, but when that temperature is reached the whole of the amber becomes a single tenaceous mass which the stirrer can hardly move. It is thus somewhat superfluous to attempt to use it.

Dissolving the Distilled Resins.

After distillation for a sufficient time the resins dissolve fairly easily in solvents, but there are difficulties in the way of getting the solutions perfectly clear. In many cases an apparently finished solution contains appreciable quantities of resin which are not really dissolved but simply swollen up. Nothing is gained by grinding the resin beforehand, for then it swells up all the more and forms a tough mass which only comes into surface contact with the solvent.

To secure proper contact between the resin and the solvent, both must be kept in constant movement. To do this many factories use the shaking machine represented in Fig. 10.

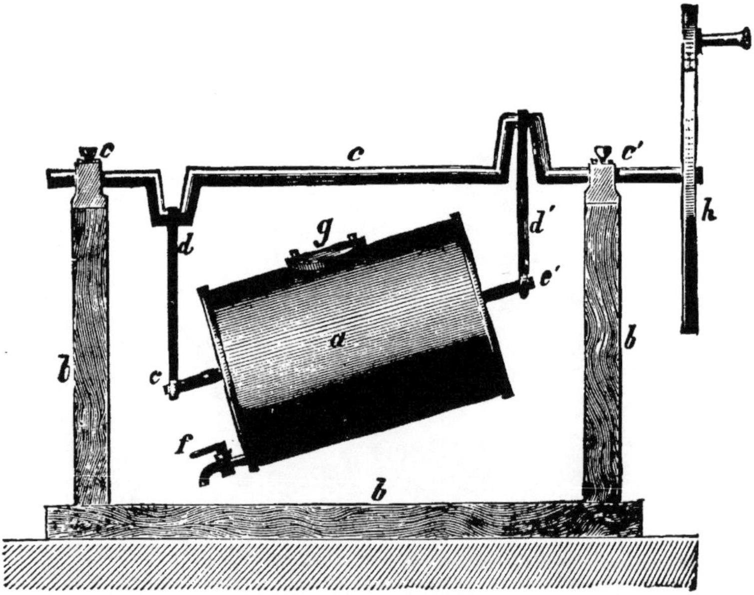

Fig. 10.

The frame *bb* supports the horizontal axle *c*, which can be rotated by means of the flywheel *h*. This axle has two cranks *d* and d^1 set 180° apart. The cranks support the vessel *a* in a slanting position. This vessel is fitted through the opening *g* with broken resin and solvent till it is two-thirds full. The wheel is then turned so as to lift first one end of *a* and then the other, so that its contents are being continually thrown from one end to the other. This brings resin and solvent into intimate contact and effects rapid solution. After a time a sample is drawn off into a glass by the tap *f*. When the sample is a thick slightly opalescent liquid, the process may be regarded as finished and the vessel *a* is emptied either into the filter or the clearing vat.

Although this apparatus answers fairly well it absorbs a great deal of power, the whole weight of *a* and its contents having to be lifted about. This serious drawback can be avoided by using the apparatus shown in Fig. 11.

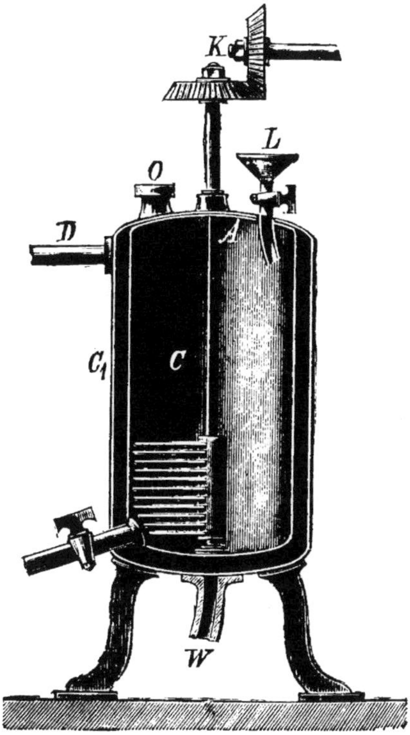

FIG. 11.

The cylindrical dissolving chamber C is jacketed with another cylinder C_1, the jacket space being a few inches wide. Steam is admitted to the jacket by the pipe D, and the steam and condensed water escape through W. The upper end of C is provided with an opening O for putting in the resin. The solvent is run in from the funnel L. In the middle of C there is a vertical axis A turned by the two bevel wheels at K. This axis carries a number of horizontal rods arranged spirally round it.

Having put in the solvent the stirrer is set in motion and the powdered resin introduced in a thin stream. When it is all in the opening O is screwed up and steam admitted to the jacket. The stirrer prevents any resin from sinking to the bottom and the joint action of the stirring and the heat finish the solution very quickly.

As soon as a sample shows a satisfactory appearance the steam is shut off and O is opened and C is emptied. This description will show that the apparatus requires only sufficient power to keep the stirrer in constant motion and this is not great even when the apparatus is on a very large scale. To dissolve the resins in volatile solvents, such as oil of turpentine for example, the apparatus shown in Fig. 12 can be employed.

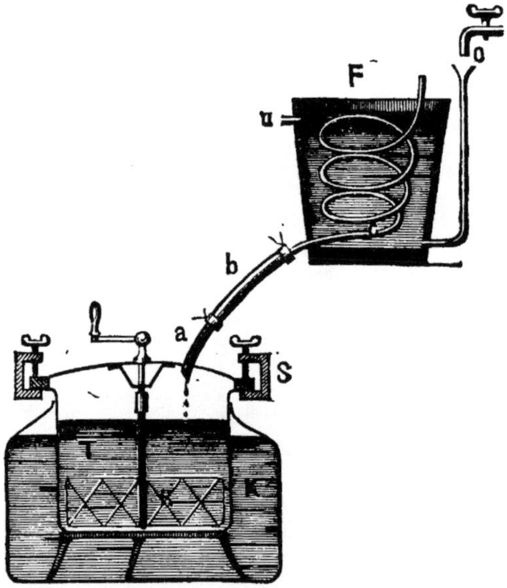

FIG. 12.

The vessel K is filled with a solution of common salt in which the dissolving chamber T dips. In T is the stirrer R, and its cover is fastened on air-tight with the clamps S. Through the cover passes the end *a b* of the worm in F. The worm opens freely at its upper end and is kept surrounded with cold water.

The resin and turpentine are heated together by boiling the salt solution which has a higher boiling point than water. The heat is assisted by the motion of the stirrer. The turpentine escaping as vapour from T is condensed in the worm and flows back again. Solution is still more rapid if paraffin is used instead of brine. The temperature can then be brought to 160° C. the boiling point of oil of turpentine.

Filtering the Resin Solutions.

As already mentioned we may consider solution as complete as soon as a sample shows that the dissolving vessel contains a slightly opalescent liquid. But a simple experiment will show that the solution is anything but complete. Under the most favourable circumstances we have a solution in which a certain amount of the resin is swimming about in a very swollen up condition. If we put the solution into a tall jar of white glass, stopper it up and put it in a place where it cannot get shaken, we shall see after several days that the upper part of the liquid is quite clear, but that below the clear part it gets more turbid the lower in the glass it is. After a few weeks the largest part of the solution has become quite clear and transparent, and a slimy tenacious mass will be seen adhering to the bottom. This consists of undissolved but highly swollen resin. Many manufacturers clear their solutions by decanting the clear liquid obtained by standing as completely as possible from the sediment. The latter is shaken up with some of the solvent and used in preparing the next lot of solution.

It is easy to see that this tedious and troublesome method is utterly unsuitable for working on a large scale, although it may answer well enough in the manufacture of small quantities at a time. It takes much too long. We can, however, get quite clear solutions very much more quickly by using filters so arranged that all undissolved particles are kept back and nothing runs through the filter but actual solution.

We know that there are a large number of different filter-arrangements which leave nothing to be desired either as regards rapidity of filtration and clearness of the filtrate, but they all have the disadvantage that they work faultlessly for a short time only. They soon begin to filter slowly and in time cease altogether to do so. The reason of this is that the slimy and swollen particles of resin block up the pores of the filtering material so completely that even very high pressure is insufficient to force the solution proper through the mass. The filter must therefore be so designed that this stoppage of the pores is prevented. An apparatus very suitable for this purpose, and one which will give large quantities of clear filtrate in a short time is shown in Fig. 13.

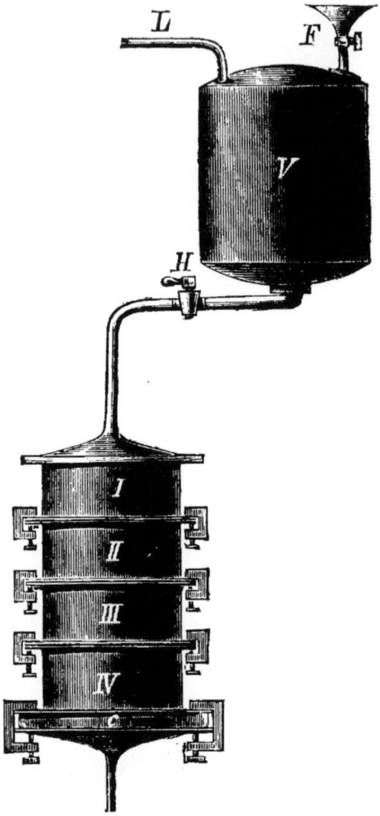

FIG. 13.

The apparatus consists of the cylinders I., II., III., IV., made of strong sheet copper varnished inside with amber-varnish. These cylinders are about eight inches high, and are connected by their flanges with the clamps as shown, thus forming a single air-tight chamber. To the lowest cylinder IV., is fastened a funnel with an outlet tube, and the uppermost cylinder I. has a funnel shaped lid, the apex of which receives a tube from the copper cylinder V. containing the liquid to be filtered. The only other openings in **V.** are one to receive the funnel F for filling it, and one for the pipe L through which air can be pumped into V. The tube to the filter can be closed by the cock H, and the funnel F can also be shut off with a tap. A rather closely woven cloth is stretched between I. and II. which are then clamped together by their flanges. III. is then clamped on in the same way and then IV. Between IV. and the funnel C a plate of cellulose about four-fifths of an inch thick is put between two cloths. As soon as the apparatus is all put together the cock H is opened and the liquid is allowed to run through the filter. As the liquid, to reach the cellulose plate, must pass through the three closely woven cloths which are stretched between the flanges of the cylinders the coarser solid particles collect on the uppermost cloth and finer particles on those below it, while only a nearly clear solution reaches the cellulose layer.

As, however, even with this arrangement the action will soon become slow, the pressure of the air in it is gradually raised as the filtration proceeds, by means of the condensing pump attached to L. The pressure must not be suddenly increased, as the filter cloths would then probably be torn. A pressure of five or even six atmospheres is often finally attained.

Care must be taken not to increase the pressure needlessly. As long as the filter is acting well, the pump is not worked, but as soon as the issuing stream shows that the filter is getting choked, the pressure is raised until the former rate of filtration is restored.

When at the close of the filtration the apparatus is taken to pieces, the surface of the cloth is found to be covered with a rather thick layer of a gelatinous mass, which consists of swollen resin. To recover this and to clear the cloths they are rinsed in the solvent used for making the solution. When, too, the pores of the cellulose plate are choked up it must be cleaned in the same way.

In this way perfectly clear solutions of amber or copal are obtained, and if oil of turpentine is the solvent employed, these can be used at once as varnishes. The coats will appear quite colourless and have a superior lustre. If the solutions are mixed with linseed oil made drying by the cold process with manganese borate, the resulting varnishes are not darker than a pale yellow, and are a great improvement on those made by dissolving run copal or amber direct in boiling linseed oil. Varnishes made by this latter method have a distinct dark brown colour, even in thin coats, and always appear of a yellowish colour on a white ground, even when laid on very thinly.

V. THE MANUFACTURE OF ILLUMINATING GAS FROM ROSIN.

If rosin is exposed to dry distillation, the decomposition takes a course which depends upon whether it is brought suddenly to a very high temperature, or a moderate heat is gradually attained. In the latter case a large quantity of product is obtained which readily condenses to an oily mass, the well-known rosin oil. The higher the temperature the darker in colour and the richer in carbon does the residue become, and finally no oil passes over. The heating is then stopped and the residual pitch removed from the still. It is true that even in this distillation of resin at moderate temperatures there is a certain amount of inflammable gas produced, but this has only very little illuminating power and is only utilisable as fuel. This gas is burnt under the still so as to economise the fuel.

When the distillation is carried out in the way first mentioned, with a very large and rapid rise of temperature, a very large amount of gas having great illuminating power is obtained. The amount of resin oil produced is very small, and very much less pitch is produced. This takes the form of a black shiny coke-like mass.

As regards the relative cost of coal gas and rosin gas, the use of the latter is distinctly cheaper for various reasons when both are made on a small scale. As the candle-power of rosin gas is double that of coal gas, only half as much of it as of the coal gas need be used. Moreover, the value of the rosin oil obtained in making the gas, although it is not in large quantities, must be taken into account. Finally, it is far easier and less expensive to purify rosin gas than coal gas. Hence the plant for producing rosin gas takes up very little room.

In those factories where rosin is used for the manufacture of rosin oil, cart grease, etc., hardly any other means of lighting is used other than rosin gas, and many examples show that even whole towns may be profitably lighted with it.

There are several apparatuses for making rosin gas, and most of them are like those used for coal gas, naturally with certain modifications. Fig. 14 shows a rosin gas-making plant in section.

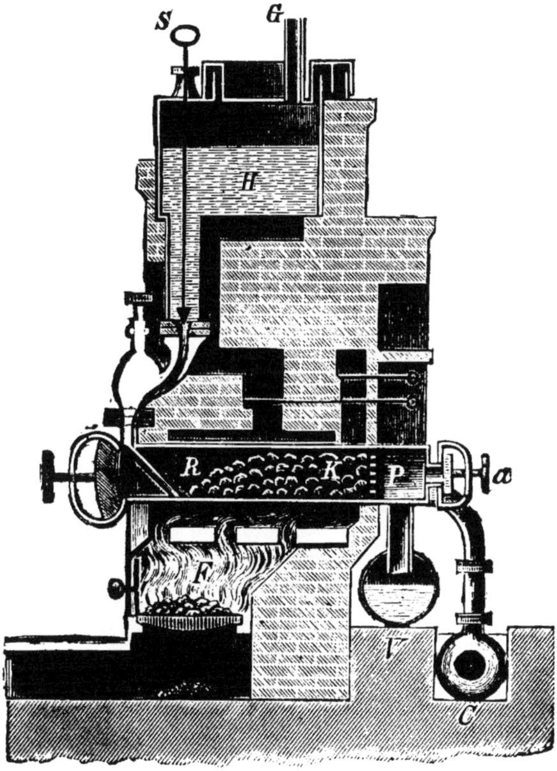

FIG. 14.

The vessel H which contains the rosin to be gasified has a cover closed with a water joint, and provided with a pipe for the escape of vapour from the heated rosin. The deepest part of H consists of a wide tube enveloped by the combustion gases from the fire which heats not only it but the rest of the bottom of H. A conical opening at the bottom of this tube has an iron cone attached to the rod S fitting into it, so that by raising the rod the opening may be partly or fully opened. Below it a pipe leads into the retort R, which is kept red hot by the fire F. The retort is of cast iron, and differs little in form from those used for coal gas manufacture. At the place where the melted rosin flows from H into the retort there is a triangular slanting plate, the object of whch is to spread out the stream of rosin falling upon it. The body of the retort as far as the perforated plate P is filled with coke. The space behind P is empty and is connected with the condenser V, where the less volatile products of the distillation collect. The pipe C which lies in water carries the gas to the scrubbers. The opening *a* is used for cleaning out the retort.

The above apparatus is somewhat expensive, and the expense is of course greater when a number of retorts have to be used, as is the case if very large quantities of rosin gas are required. There is, however, a much simpler form of apparatus which can be made large enough to furnish very considerable quantities of gas. The saving is effected by doing away with the furnace and the retorts. Such an apparatus is shown in section in Fig. 15.

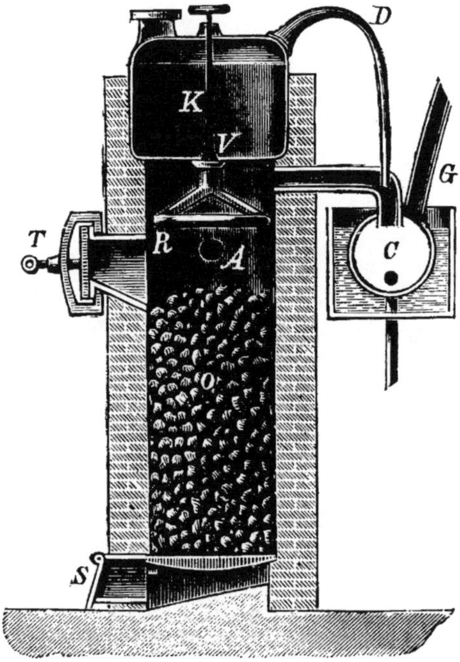

FIG. 15.

The closed boiler K stands on the top of a cylinder of fire-brick, and receives the rosin to be gasified. It is provided with a manhole through which it is charged, and the fumes escaping from it pass by the pipe D into the condenser, together with the gas from the fire-brick cylinder. The condenser C is a wide iron pipe lying horizontally in water. By means of a pipe the liquid products which condense in it can be drawn off, while the gas passes by another tube at the opposite end of C to the scrubbers.

In the side of the brick cylinder there is an opening T which can be closed gastight by an iron plate. A grate is put at the bottom of the shaft, and the space below it can be reached by an opening in the side. This opening is lined with an obliquely cut iron pipe, and closed by the heavy iron plate S. This plate can be moved sideways, so as to open a slot of any desired width. A short pipe in the bottom of K is closed by the valve V, which can be screwed up by the rod, so as to allow the rosin to flow out fast or slow as may be required. The part of the tube below the valve has three or four branches, which lead into a hollow ring with many perforations on its under side. Hence when V is lifted a large number of the smal streams of melted rosin flow down into the brick shaft or stove.

When the apparatus is to be set in action the shaft is filled with coke through T. T is then shut and by means of a damper the pipe A is put into communication with a chimney. Fire is then put below the grate through S. When the whole of the coke is at a full red heat, A is closed, so that all vapours escaping from the stove must pass into the condenser C. The valve V is then opened, so as to drip the melted rosin on to the hot coke. The rosin is decomposed, and the liquid products condense in C, while the gas passes by the pipe G to the scrubbers. The draught is regulated by S so as to keep the coke glowing, and the supply of rosin is managed by adjusting the valve V in such a way that the production of gas proceeds at a uniform rate. The process must not be so driven that the condensation of the rosin oil in C is incomplete. The supply of air through the narrow slot in S is enough to keep the coke in the stove O red hot. As it passes up through the coke, carbon monoxide is formed and becomes mixed with the rosin gas. As carbon monoxide is itself combustible, and there is not enough of it appreciably to affect the illuminating power of the resin gas, it is unnecessary to separate it.

Purification of Rosin Gas.

The gas escaping from C must be purified before it goes to the gasometer, but as the gas is already very pure this can be done in a very simple fashion. The gas is free from compounds of sulphur, and the chief impurities are carbon dioxide, and also usually small quantities of the vapour of acetic acid, which escapes from the condenser by reason of its very volatile nature. It is absolutely necessary to remove the carbon dioxide which much lessens the candle power of the gas. The acetic acid is not objectionable on this account, but may in time corrode the iron distribution pipes through which the gas is led to the burners. Both impurities can be removed simultaneously by passing the gas over lime slaked with just enough water to form a dry powder. This absorbs both mpurities. The slaked lime is put into shallow iron pans, the covers of which are closed by a water joint. These pans contain iron shelves close together and the lime is spread in a thin layer upon the shelves. These are so arranged that the gas must pass over each, travelling the length of the pan as many times as there are shelves. The gas then passes direct into the gasometer, whence it passes to the burners in the usual manner.

As already stated rosin gas is much superior in lighting power to the best coal gas, on account of its high content of carbon. This peculiarity of its composition makes it inadvisable to burn it from an ordinary fish-tail burner; an Argand or other lamp, with which it can be mixed with air before combustion, should be used. The best way of all of using it is with a mantle, whereby we get the most light with the smallest consumption of gas. The high percentage of hydro-carbon in rosin gas makes it burn, when mixed with enough air to form a blue flame, with great evolution of heat, and it forms therefore an excellent fuel for heating stoves or soldering irons.

The yield of rosin in gas compares very favourably with that of coal and other gas-forming materials. Pure American rosin will give 1,600 to 1,700 cubic feet of gas per cwt. As we always get some rosin oil in making rosin gas, the oil will pay part of the cost of manufacture, and the cost is easily reckoned for any given price of rosin.

Rosin has also been used to make coal-dust suitable for gas-making, by stirring the dust into melted rosin till the mass is too thick to be stirred any longer. The mass is broken up when cold and set, and charged into the retorts either alone or mixed with lumps of coal. Formerly coal tar was used instead of rosin for this purpose, but a far more remunerative use for coal tar is now found in the manufacture of dyes and other well-known products. The cheapest rosin is naturally used for mixing with the coal dust.

The Dry Distillation of Rosin.

VI. ROSIN OILS.

When any form of resin from which the essential oil has been removed is heated in closed vessels, it begins to decompose at a temperature depending upon the nature of the resin, and yields a large number of products of decomposition, partly gaseous and partly liquid. The final residue is a shining, black, scoriaceous coke.

The gaseous products so obtained are for the most part combustible, and burn with very luminous flames. The products condensible into liquids from the vapours escaping from the stil are very various in composition. When allowed to stand, the liquid distillate soon separates into two sharply-marked layers—a watery layer and an oily layer. The latter constitutes the crude resin oil. This consists of a large number of distinct hydrocarbons, and its composition depends not only upon the kind of resin treated, but upon the way in which the process is conducted. If the temperature is raised very gradually the products are quite different, especially from a quantitative point of view, from those got when a high temperature is reached as quickly as possible.

The commercial value of the resin oils varies enormously with the sort of resin. For example, the oils obtained from certain hard resins such as amber and copal, serve very well for solvents of the same resin in varnish-making, but the amount of oil which such resin yields is rather limited. On the other hand there is a large yield from the commoner resins, and especially from those obtained from coniferous trees. As is well known we have vast quantities of this resin at our disposal, and it was used on a fairly arge scale in various industries as early as the middle of the nineteenth century. In 1835 the French chemist, Fremy, one of the first to study the products of the dry distillation of resins with thoroughness, prophesied that the products from rosin would become of great commercial importance, a prophecy which has been strikingly fulfilled. As stated, the products obtainable by the dry distillation of rosin were exciting great attention in Europe at the beginning of the second half of the nineteenth century, and factories of considerable size were erected for this branch of industry.

Nevertheless, the real start of the rosin oil industry dates from the time when the Americans began to exploit the product of their vast pine forests in a rational manner, and to make oil of turpentine and rosin. As the preparation of the crude rosin is carried out in all its stages in a practical manner in America, the quality of their rosins, even that of the second-rate brands, is very satisfactory, and they are all very suitable for the manufacture of rosin oil.

Although very many of the European manufacturers who distil rosin oils use American rosin for the most part, it must not be forgotten that Europe produces very important quantities of rosin, which can compete on an equal footing with the American product, if it is obtained in an equally rational manner. In the mountainous districts of Europe, especially in the Pyrenees and the Carpathians, we have large pine-woods, the yield of which might be considerably increased by working up the crude rosin into colophony and oil of turpentine by scientific methods. In Siberia there are enormous primeval forests largely consisting of pine trees, and there is no doubt that these forests, now opened up by the completion of the Trans-Siberian Railway, could yield enough rosin to supply all the needs of the whole of Europe.

In the times previous to the development of rosin oil manufacture, the chief object of distilling colophony was the product of illuminating gas. The tar, formed in the process, which is crude rosin oil, was considered of little value, and nearly all of it was used to eke out the fuel for heating the gas-retorts. Since, however, the tar has been scientifically investigated, we have found that it is the gas that must be considered as a by-product, and that the oil is of far greater value. A very analogous thing happened with coal distillation. For many years the black evil-smelling coal tar was regarded as a troublesome by-product attending the production of gas for lighting purposes, and there was hardly any use for it except in heating the retorts. But when we came to know more about the constituents of coal tar we arrived at the epoch-making discovery of the manufacture of dyes from it, to say nothing of other valuable products, and the whole state of affairs was changed in a twinkling. In a few years the hitherto practically worthless tar had become a highly prized raw material, as did also the gas liquor. The gas retired into the position of a by-product, and the uses now made of coal tar afford a most interesting example of exhaustive exploitation of a raw material.

Although the products of the dry distillation of rosin have not found such comprehensive and vast fields of usefulness as those obtained from coal, they do nevertheless find so many uses that it can be fairly said that the distillation gives no product which is destitute of value. No better proof of the scale on which this industry has developed can be given than the statistics of the imports of American rosin into Europe. The quantities imported towards the end of the nineties already amounted to more than a million barrels, and these quantities have since considerably increased.

The following is a list of the products obtainable by the dry distillation of rosin (*a*) gas for lighting and heating, (*b*) acetic acid from the watery part of the distillate, (*c*) light volatile oils, which are not only excellent illuminants but valuable as solvents for many substances, (*d*) heavy oils largely used in the manufacture of lubricants for vehicles and machinery. The whole business of cart-grease-making as now practised depends upon proper treatment of these heavy oils, (*e*) driers. Proper chemical treatment of rosin oils will convert them into liquids which can be used in the mixing of paints exactly like the drying oils, and will give a mixed paint quite as good as one prepared with boiled oil, and at a much smaller cost. (*f*) Lampblack. A lampblack can be made from rosin oil, and is one of the finest lampblacks that can be made in respect both of minute subdivision and of beauty. The purest sorts of it are used in artistic printing in copper-plate work and in lithography, and also for the preparation of an Indian ink which is equal to the best Chinese make.

THE MANUFACTURE OF ROSIN OILS ON A LARGE SCALE.

If we put rosin into an ordinary still connected with a worm, such an apparatus as used to be employed for distilling brandy, we are in a position to undertake the dry distillation of rosin. The oldest piece of apparatus used for the purpose differ from brandy-distilling plant simply in their great size. Some stills have been built for treating a ton of rosin at a time, that is to say, big enough to hold four tons; a still must not be filled with cold rosin to an extent exceeding one quarter of its capacity as the rosin swells and froths up enormously on heating, These stills were invariably heated over a naked fire, and had all the disadvantages of apparatus so heated. Of these the most serious is that it is impossible to regulate the temperature and it is especially necessary in the distillation of rosin to do this. We must be able to keep the temperature steady at any given point. If this is not done the decomposition of the rosin will be irregular and spasmodic, and will not follow a uniform course. If it is necessary to use a naked fire the distillation must be ceaselessly watched from first to last, and the fire must be tended with the greatest care. If the temperature should happen to go up suddenly the probability is that the development of vapour will be so rapid that the condenser will be unable to cope with it, so that quantities of resin oil will escape uncondensed and be lost. There is also the danger with such rapid generation of gas of bursting the still. The temperatures required for the distillation of resin rise to a fairly high figure. Beginning with a cistillate coming over at about 130° C., the heat is gradually indreased to 350 or 360 deg. This finishes the distillation and leaves nothing in the still but black pitch in a liquid state.

Plant for Distilling Rosin.

In their essential features rosin distilling apparatuses can be divided into two classes, those which can be worked uninterruptedly, and those which work at intervals. The former kind is the only really suitable one and can be subdivided with reference to the method of heating into those with direct and those with indirect firing. It has already been mentioned that formerly an ordinary still with a movable head and a worm was used for rosin distillation. With an apparatus of this sort, as soon as the distillation is over and the pitch has been run out, the still must be allowed to cool before recharging. Moreover, the valuable uncondensable gases which are evolved were totally lost. The large consumption of fuel, the long and constant interruptions between distillations and the uncertainty attending the whole course of the process have finally brought it about that such a plant is now hardly ever used. It is therefore sufficient to allude to it. It is. of course of historical interest that it was the first apparatus used for the purpose.

Much more suitable, and hence now almost exclusively employed n modern well-appointed factories are the apparatuses in which superheated steam is used for heating instead of an open fire. The use of superheated steam enables us to regulate the temperature of the still with perfect accuracy, with the result that the yield of the valuable products of the distillation is materially increased as compared with that resulting from direct firing. It has been observed that to obtain the best yield of the most valuable among the distillation products of rosin, it is necessary to get the vapours from the still into the condenser as quickly as possible, to prevent them from being further decomposed by the hot sides of the still. To effect this an arrangement is adopted in which the vapours are sucked out of the still by means of a pump as fast as they are formed. In the combination of heating by superheated steam and the use of such a pump we have the best apparatus for the distillation of rosin and we proceed now to describe one of that kind constructed by the author.

Apparatus for Superheating Steam.

Every apparatus for superheating steam must consist of a boiler and a superheater. The boiler is an ordinary one capable of supplying steam at a pressure of a few atmospheres. The superheater may be constructed in various ways. Figs. 16 and 17 show one of the older forms used in connection with a large plant.

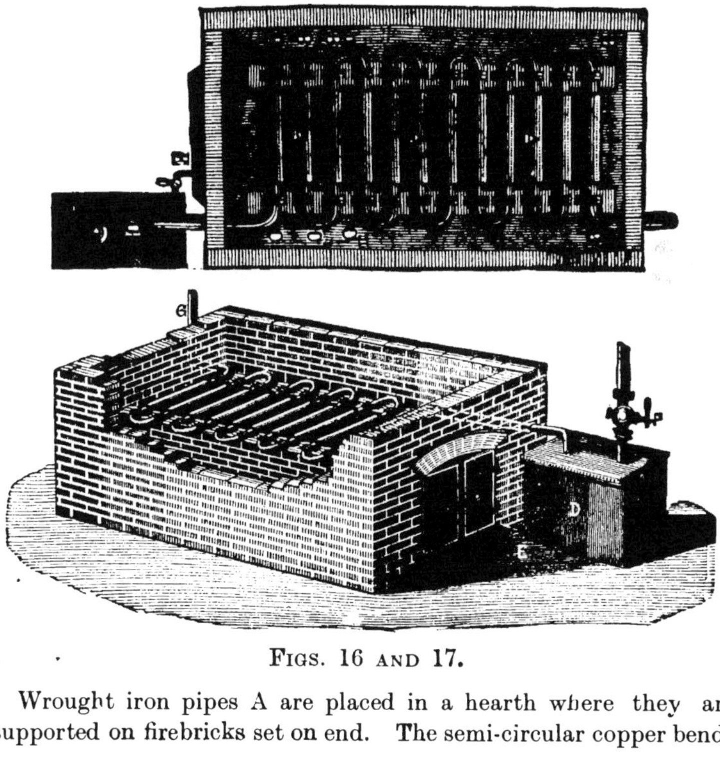

Figs. 16 and 17.

Wrought iron pipes A are placed in a hearth where they are supported on firebricks set on end. The semi-circular copper bends

B and C unites these pipes into a coil. In front of the hearth is an iron chamber D into which the steam pipe from the boiler opens. The object of D is to catch any water brought over by the boiler priming. A fire is made in the hearth so as to heat either all or part of the piping according to the amount of superheated steam required. The opening of the cock F admits the steam from D to the coil which can be heated to a temperature of 400° C, or upwards, without increasing the steam pressure. At G where the steam leaves the superheater a thermometer can be used to enable the temperature to be regulated. This thermometer is, however, rarely used as the attendant soon learns how to fire so as to get steam of the temperature required.

Another superheater, also of great excellence, is shown in Figs. 18 and 19.

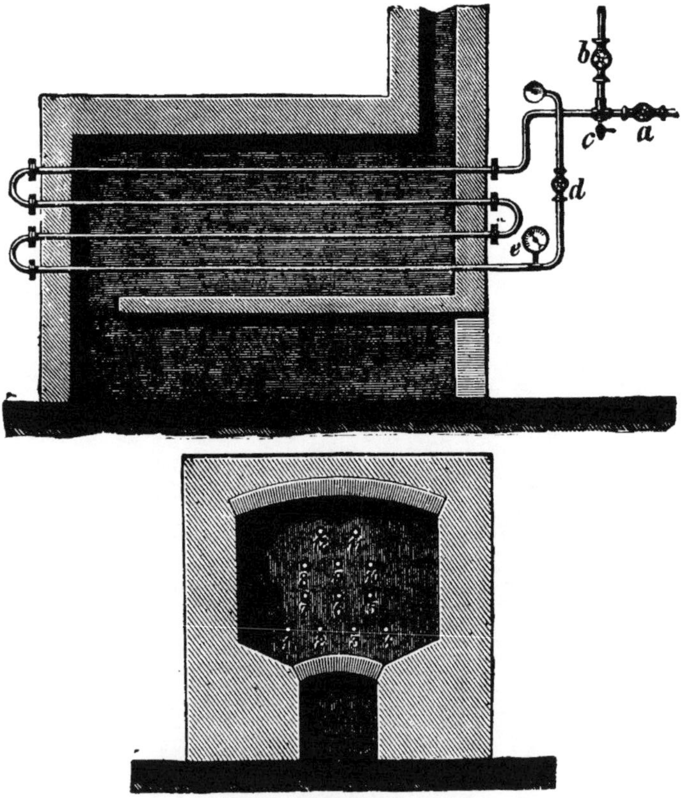

FIGS. 18 AND 19.

Here the coil is made of drawn wrought iron tubing. It lies in four layers, four tubes in the lower set, three in each of the next two, and two only in the upper set. The tubes are joined by elbows placed outside the furnace.

In using superheated steam we are in a position to carry out with only one still a process which only needs to be interrupted by short intervals, sufficient for the removal of the pitch and for recharging the still with rosin, but for the sake of economising fuel and with a view to the fact that when there is only one still any accidental damage to it stops the whole work of the factory till it is repaired or replaced, it is best to work with an apparatus in which two stills are combined with the same superheater. This arrangement has the additional advantage that the superheater can be kept always heated, for towards the end of the distillation the temperature of the still must be very high and even at the beginning plenty of heat is wanted to fuse the new charge of rosin and begin the distillation as rapidly as possible.

To heat the two stills independently the pipe by which the steam leaves the superheater is divided into two branches, one for each still. Each branch is provided with a tap so that steam can be sent to one still only or to both, and the amount of it can also be regulated.

In order to ensure that the largest possible proportion of the heat of the steam should be utilised in heating the stills, it is always made to pass through a flat coil just above the bottom of the still. The steam passes from the coil into the space between the still and the brickwork in which it is set, so that it heats the still from the outside as well as from the inside.

These are the general principles of proper construction of plant for distilling rosin oil and we shall again refer to the most important types now made.

Combined Distilling Rectifying Apparatus.

As the greatest saving of time and fuel is secured by combining the apparatus for producing the crude rosin oil with that for rectifying it in such a way that the same superheater can be used for both, we have constructed a plant which undoubtedly works more cheaply and quickly than any other. As it also provides for the complete collection of the considerable amounts of gas invariably produced in rosin distillation we can justly assert that the apparatus as constructed by us answers fully every requirement which can be made of a plant of the kind. Our apparatus consists of the following parts :—

1. A boiler large enough to give an uninterrupted supply of

steam which need not, however, be at a very high pressure, one of two or three atmospheres being quite enough.

2. A superheater built on the plan shown in Figs. 18 and 19. The superheater ought to be fairly large, for its heat is taken most advantage of when long tubes are used, as they deliver the steam at a very high temperature such as is required for resin distillation and rectification.

3. Two exactly similar stills, each provided with a charge heater for the resin. Each still has a worm which is connected to an apparatus where the gases are separated from the liquid products.

4. A rectifying apparatus large enough to rectify quickly the combined output in crude oil of the two stills so that there is no need for any interruption of the work.

Figs. 20 to 23 inclusive show the separate parts of the whole plant.

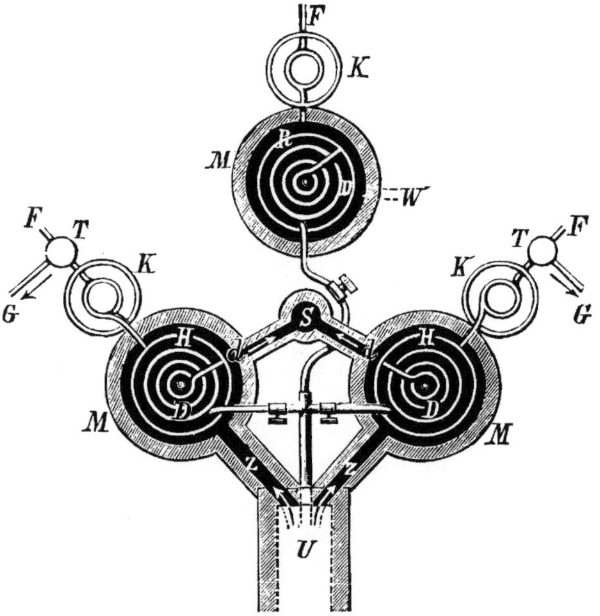

FIG. 20.

Fig. 20 shows the arrangement of the two stills with their condensers and the vessel in which the gases and the liquids are separated and also the rectifying apparatus. The steam pipes from the superheater U enter the still H, H, in which they traverse the coils D, D, the steam finally escaping at *d*, *d*. The stills are surrounded by brick chambers M into which come the hot gases from the fire of the superheater. These gases together with the steam coming from the coils pass over the outsides of the stills and finally escape by the chimney S. This arrangement utilises the heat of the superheater in the most complete manner possible. The rectifier R is put in front of the stills, is like them surrounded by a steam jacket and provided with an internal coil fed with steam from the superheater. This steam becomes fully condensed to water which flows away from the bottom of the jacket by a pipe.

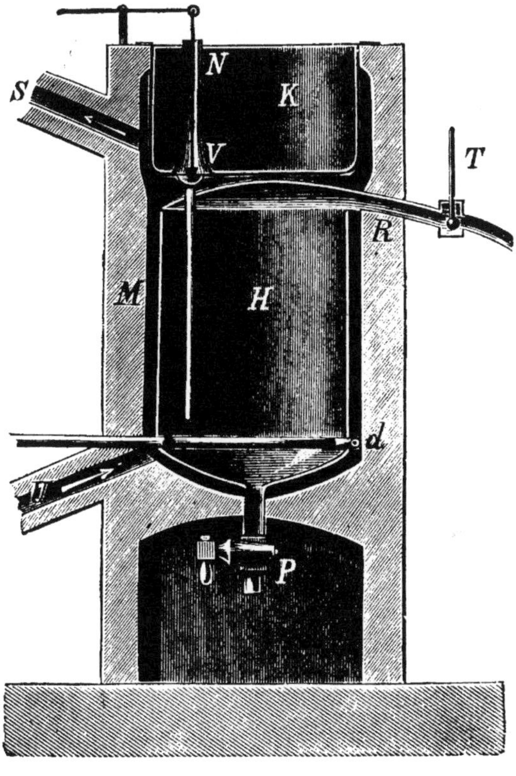

FIG. 21.

Fig. 21 shows a vertical section of a rosin distilling apparatus. The coil of the still opens freely at *d* into the space between the still and the brickwork M. The pipe R for the exit from the still of the distillation products is at the upper part of the still. A wide part of R receives the bulb of the thermometer T, the readings of which show the progress of the distillation. At the lowest part of the still is a pipe P with a cock, so as to run out the pitch when the process is finished. Immediately above the still is the charge heating vessel K, heated like the still by the fire gases from the superheater, which enter at U. The fire gas leaves for the chimney by S. The heater K has a valve V which can be raised to admit molten rosin to the still. K is made large enough to hold the rosin required for one distillation. As soon as it is over and the pitch is run out, the charge is run from K into H by lifting **V** and the distillation is at once re-commenced. The pipe from K goes nearly to the bottom of H. To prevent pieces of wood and other mechanical impurities from entering the still the valve V is completely enclosed in the tube N made of wire gauze acting as a filter.

In rosin distillation liquids condensed in the worm are obtained as well as uncondensible gases. To separate these the apparatus shown in Fig. 22 in vertical section is used.

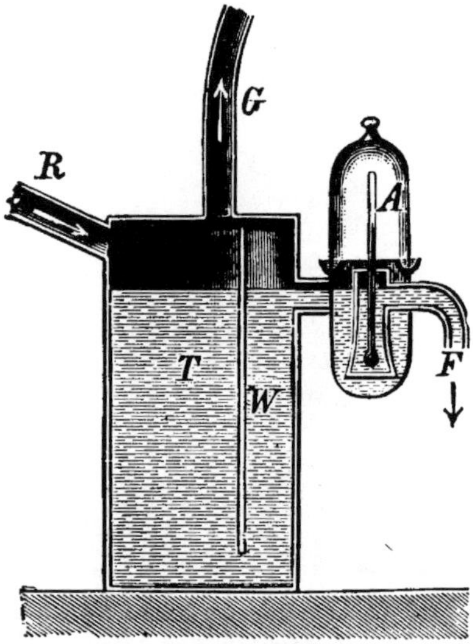

FIG. 22.

The separator T is so connected with the cooling worm that the whole of the products of the distillation must enter it. The cylindrical vessel T is divided by the partition W, which nearly reaches the bottom of the cylinder, into two unequal parts. The products from the still come in by R and the liquid portions of it collect in T and finally fill it. There is then an overflow in an unbroken stream at F, while the gases escape by the pipe G. This may be made to lead into a gasometer, and the gas can then be used to light the factory. Any gas over and above what is required for this may be burnt in the superheater to eke out the fuel there used. A is a hydrometer which shows the specific gravity of the liquid ssuing from the separator.

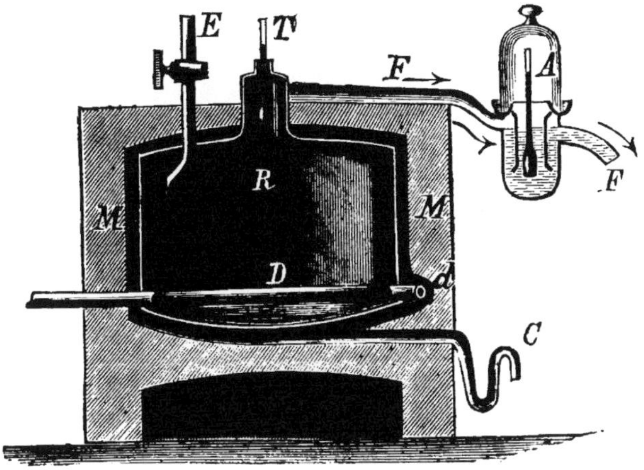

FIG 23.

The rectification apparatus is shown in Fig. 23 and consists of a still R from the dome of which the pipe F carries the vapours to the worm. The thermometer T in the dome shows the temperature of the issuing vapours, and the measuring vessel fixed at the lower end of the worm contains the hydrometer A for determining the specific gravity of the distillate. E is the pipe which brings the oil to be rectified into the still. D is the steam coil which opens freely at *d* between the still and its setting. C is the pipe for the escape of the condenseu water. It is not difficult to arrange matters so that the rectification apparatus is heated by burning the gaseous products of the distillation of tbe resin. In this case a larger gasometer must be used and burners which mix the gas with the necessary amount of air for its proper combustion. Unmixed with air the resin gas burns with a very smoky flame on account of the high percentage of carbon in it, but mixed with the proper volume of air it burns with a very hot and non-smoky flame. This method of heating is to be recommended as it admits of accurate regulation of the temperature.

A great variety of types of apparatus for distilling rosin have been proposed and the attention of practical men has been particularly drawn to that of Kramer and Flammer. This apparatus does not differ in its general arrangement from ordinary distilling plant. It is a vacuum apparatus and the vacuum is produced by a steam injector which is connected to the pipe from the still and sucks out the vapours as they are generated. This only secures a moderate advantage, for all air is in any case driven out by the time the distillation has been in progress for a short time by the gases and vapours which are evolved from the resin. There is, however, a distinct advantage in the use of the injector when the distillation is effected over a naked fire. It is then impossible to keep the temperature within bounds and the sides of the still become much too hot, with the result that the decomposition often proceeds too far and the rosin oil is to a large extent converted into gaseous compounds. The rapid extraction of the vapours from the still when the injector is used prevents this injurious action. When the heating is done with superheated steam, a very little care will enable the temperature to be very closely regulated by controlling the flow of steam, and no overheating of the sides of the still is possible. In this case the sucking out of the vapours seems entirely supererogatory.

VACUUM APPARATUS.

The Kramer and Flammer apparatus for the distillation of rosin in a space free from air is claimed by them to give a rosin oil superior in quality to that furnished by other apparatuses. The construction of the apparatus is shown in Fig. 24.

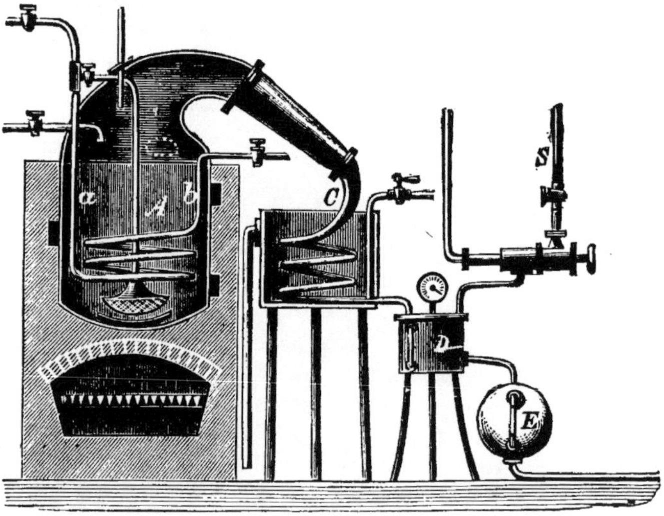

FIG. 24.

The still A is made to hold up to five tons of rosin. It is set in brickwork and heated over an open fire and also by steam circulating in the tube *a b*. The products of distillation pass into the cooling worm C, whence they pass into the air-tight receivers D and E. The injector S communicates with D and continually sucks the vapours out of the still. To prevent waste of the combustible gases coming from the still it would seem advisable to lead the outlet pipe from the injector into the fire.

In order to ascertain what difference there is between distillation without air and distillation as ordinarily conducted we have instituted an accurate series of experiments. Before beginning the distillation we expelled all the air from the apparatus with a stream of carbonic acid and then kept that gas flowing continually all through the distillation so that it also swept out quickly the products of the heating. As the presence of oxygen which is supposed to have an injurious action on rosin oil was excluded by the continual stream of carbonic acid, the conditions were exactly the same as in distilling in a vacuum. Accurate observations could, however, detect no difference, either in the amount or in the quality and nature of the products, whether the distillation was done in carbonic acid over a naked fire or by superheated steam without carbonic acid. This result was sufficient to make us discontinue the use of the injector. As already stated all the air in the apparatus is driven out very early in the distillation and the whole of the interior of the apparatus not occupied by melted resin becomes taken up by rosin-gas and the vapours of the rosin oil. In this way the action of atmospheric oxygen on the oil is prevented.

THE PROCESS OF RESIN DISTILLATION.

In the old factories which practised the old crude method, and where the whole apparatus consisted of a single still heated over a naked fire, and connected with a worm, the still was charged to at most two-thirds of its capacity with rosin. It was then heated, and the fire was gradually increased until a drumming noise within the still announced the beginning of the decomposition of the rosin and drops of distillate began to issue from the worm. Then for a long time the attendant sought so to manage the fire that the distillate flowed at a uniform rate, but towards the end, when the distillate flowed more slowly, the fire was urged strongly and kept at its utmost height till all distillation of liquid ceased. The fire was then drawn, the pitch run out, and the still recharged.

Independently of the enormous waste of fuel which seems quite unavoidable with this primitive method, the combustible gases from the resin being usually allowed to escape unburnt, the yield of the more valuable products of the distillation is much less than it should be, because the rosin being a very bad conductor of heat, those parts of it in immediate contact with the sides of the still get overheated, so that far too little oil and far too much gas is produced, while the oil actually obtained is of inferior quality. As in the present state of the industry it would hardly occur to anyone to distil rosin over an open fire, we will leave that method and proceed to describe the course taken when the distillation is executed with superheated steam. It must always be borne in mind that the aim of the foreman is to prevent the temperature of the still from becoming higher than is absolutely necessary for proper and steady work. Great care must be taken that the process never becomes violent. If it does it is a sure sign that the temperature has become excessive and the steam must be at once, wholly shut off until the flow of the distillate has greatly diminished and no noise can be heard in the still. The superheated steam is then again admitted, but cautiously, and the distillation is brought back to its normal rate of progress.

The first thing to be done is to fill the charge heater over the still with resin and then turn on the steam. This warms up the whole apparatus and the rosin in the charge heater is soon melted. It is then run into the still and the charge heater is refilled ready for the next distillation. The first time a distillation is effected the time required for the proper performance of each of the various steps should be noted. When this has been done the attendant in charge can be furnished with the data, and this, together with practical experience of the mixture in the still and the rate at which the liquid distillate leaves the worm, will enable him to manage the process properly.

The beginning of the decomposition is announced by a slight rustling noise inside the still caused by the escape of bubbles of steam through the melted rosin. When the temperature of the resin has got to about 150° C. a liquid distillate begins to appear. This is turbid and consists of a mixture of various liquids which separate into two layers on standing. As long as the turbid distillate flows the temperature is kept steady, but as soon as the flow ceases, it is further raised quickly to 200° and the receiver is changed. Receptacles must be completely filled with the first distillate and left standing till the separation into two sharp layers is complete. The lower layer is then run out into a special vessel. The upper layer is a mixture of essential oils and is known as crude pinoline. The lower contains considerable quantities of acetic acid accompanied by smaller amounts of formic acid and other members of the fatty acid series. By the time the temperature has reached

200, the distillate will flow pretty evenly, and the heat is again kept steady till the flow lessens markedly. The temperature is then again raised, but this time very gradually, because experience has shown that the more slowly it rises at this stage the larger the yield of rosin oil. The quicker it rises the less oil and the more gas shall we get.

When the temperature has reached about 210° C. the flow of the distillate has become stronger, and the operator waits till it again slackens before heating further. No definite estimate of the time required before a further heating is needed. It varies with the kind of rosin under treatment, and the first distillation of a fresh consignment of rosin must always be carefully superintended to serve as a guide for the subsequent batches.

In this way the temperature is gradually raised by 10° at a time until a heat of from 350° to 360° is attained. It should never exceed 360°. It is true that by using higher temperatures a little more oil is obtained, but there is imminent risk of coking the pitch in the still. It would then adhere strongly to the steam coil, which is the hottest part of the apparatus, where being a bad conductor of heat it would practically destroy the heating power of the coil and render it imperative to stop work, open the still, and clear the coke off the pipe. If, however, the temperature of 360° is not exceeded, the pitch in the still remains a mobile liquid which runs freely from the cock. When the pitch is discharged the pitch-cock is shut, and the still is refilled with the melted rosin in the charge heater. The valve for doing this should be large so that the still is quickly filled. When the still is completely recharged, but not before, steam is carefully admitted to the coil and the distillation is soon in full swing, as the apparatus has never cooled after the previous operation.

In a properly appointed factory, the distillation apparatus is always in duplicate, and the work is so managed that one distillation begins just as the other finishes, so that one man can look after the two apparatuses, while another looks after the rectification, for which one still is sufficient. The products obtained in the crude distillation may be regarded as finished products, but as we shall see, it is now almost the universal practice to rectify the oil, and the larger factories rectify their own products.

The Yield of Crude Products in Resin Distillation.

The amount of the products obtained depends mainly upon two factors, the nature of the rosin, and the way in which the distillation is conducted. As regards the rosin two lots of commercial rosin are never exactly alike. Some of that on the market is so impure with earth and stones that it gives a decidedly low yield of distillation products. It also happens, but less commonly, that the turpentine has not been thoroughly extracted from the rosin at the place of production. Such rosin may yield considerable quantities of turpentine oil at the commencement of the distillation. This is simply to the advantage of the rosin oil distiller, as oil of turpentine is a valuable product.

Hence to obtain reliable data of the yield we must confine ourselves to rosin free from mechanical impurities. It is therefore advisable to add to the distilling apparatus one for melting and filtering the rosin.

The way in which the distillation is carried out has great influence both on the amount and on the nature of the products. If the rosin is rapidly brought to a very high temperature, we get far more gas than when the temperature is raised gradually, with intervals during which it is kept constant. It is therefore essential to bring the temperature at the beginning of the distillation to about 120° C, and keep it there as long as the distillate flows freely. At this temperature the liquid always contains certain quantities of acetic acid and water and can be used for the production of that acid. When the flow has nearly ceased, but not before, the temperature is raised to 160° C. and kept there till the flow again nearly ceases. The temperature is then again raised and, the process is continued step by step, keeping each successive temperature steady as long as it yields a fair amount of liquid distillate. Thus when a temperature of 360° has been reached, the distillation may be regarded as finished and nothing remains in the still but melted pitch.

Although the various oils in the first distillate must be separated, by a subsequent rectification, care should be taken during the first distillation to change the receiver every time the temperature is raised. In this way a good deal of separation of the oils is effected and the rectification is greatly assisted. This affords a strong reason for effecting the distillation with superheated steam, as in that way only can the temperature be regulated accurately enough to get the successive elevations at the proper intervals. How great is the influence of various methods of distillation upon the yield, is shown by the following examples of actual results obtained by various manufacturers. 500 lb. of American rosin gave :

	No. 1.	No. 2
Light resin oil and acid water	45 lb.	50 lb.
Heavy resin oil (crude distillate)	375 ,,	380 ,,
Pitch	35 ,,	40 ,,
Gases	5 ,,	? ,,

That these results are valueless appears from the circumstance that No. 2 gives 470 lb, *i.e.*, 30 lbs. less than the original rosin.

Somewhat nearer the truth are the results next to be given. 200 lb. of rosin gave :

Pinoline	6·2 lb.
Resin oil	170·2 ,,
Coke	7·8 ,,
Water and acetic acid	5·0 ,,
Gas and loss	10·8 ,,
	200 lb.

Even these results, however, appear of little value, for the yield of pinoline, *i.e.*, that part of the distillate having a specific gravity below 0·9, is strikingly small, being only 3·1 per cent.of the total rosin. The fact that the residue is given as coke and not pitch, together with the figures for gas and loss, show that the distillation was improperly conducted. The contents of the still must have been rapidly brought to a high temperature, one so high in fact that the pitch was decomposed and coked, and large quantities of gas were formed. The following results were obtained with the apparatus of Kramer and Flammer from 10,000 lbs. of rosin in round numbers :

Light rosin oil (rosin essence)	300 lb.
Pale yellow rosin oil	5000 ,,
Blue rosin oil	1700 ,,
Green rosin oil	1500 ,,

The most accurate data we possess of the yield of the various distillation products are certainly those obtained by the chemist Thenius. He gives the following average yields in percentage of the weight of rosin used :

Acid water	5.7	
Light crude rosin oil	11·4	(specific gravity 0·89)
Heavy ,, ,, ,,	50·0	(blue with specific gravity 0·93)
So-called train oil	10·4	(specific gravity 0·94)
Pitch	18·5	
Gases	4·0	

If the distillation apparatus permits the greatest possible uniformity in heating the whole mass of rosin, the quantities of the various products are entirely different, and much larger quantities of pinoline (camphene) are obtained, and this must be regarded as a better result, as the light oils are more valuable than the heavy ones.

It is well known that the resins are very bad conductors of heat, so that it is very difficult to heat large quantities of rosin in a uniform manner. In using a naked fire, the rosin next the sides of the still may be fully decomposed, when the innermost portions of the mass have hardly begun to yield any distillate. By far the easiest way of heating the rosin uniformly through and through is to use super-heated steam, so that distillation by its aid is the only really rational method.

VII. THE NATURE OF THE CRUDE PRODUCTS.

In the factories with primitive plant, in which the distillation of the resin was carried out over a naked fire, no attention was paid to the separation of the various oils according to their boiling points and specific gravities, and all of them, light and heavy alike, were collected in the same receiver. The separation was effected solely after the first distillation.

In this way a distillate was obtained varying in colour from a pale yellow to a dark brown. Experience has shown that the age of the rosin when it is distilled has a notable influence on the colour of the distillate. Old dark rosin always gives a darker oil than pale soft resin. When plant of more modern construction is used which effects some separation, if only partial, of the various oils, during the course of the first distillation the light oil comes over first. This in its crude state is pale or dark yellow. The heavy oils which follow are brown. As the distillation proceeds, the oils come over darker and darker and finally attain a very dark brown indeed.

All the crude distillates show fluorescence. The light oils have a bluish shimmer, but the heavy oils show a strong blue shimmer The fluorescence of the last distillates shows a distinctly green shade. If crude rosin oil is kept where air has access to it, even to a limited extent, the whole liquid assumes a deep blue colour, and is then a special article of commerce known as "blue oil." Some makers also distinguish the oils which show a more or less pronounced green or brownish olive hue, as "green oil" or "train oil."

Together with the liquid products, the distillation gives a large quantity of gas, which when lighted burns with a very bright light, and far exceeds coal gas in illuminating power. Hence rosin gas can be used very advantageously either for lighting purposes or as a fuel.

Before the oils proper, a watery distillate comes over which contains considerable quantities of acetic acid, and in works where it is made on a sufficiently large scale, this is used for the preparation of pure acetic acid, or of the acetates of lime or iron.

When the distillation has been properly conducted, the pitch remains in the still as a thin black liquid, which sets on cooling to a solid mass. If the bottom of the still has been overheated, part or all of the pitch is converted into a shiny black coke, much resembling ordinary coke produced from coal. This rosin coke makes excellent fuel.

The Chemical Constituents of Resin Oil.

Rosin oil has a very complicated chemical composition. Although many chemists have worked at it, we cannot as yet consider that the subject has been thoroughly investigated. All rosin oils consist of hydro-carbons belonging to the family of terpenes and also of smaller quantities of ternary compounds of the nature of aldehydes. The best investigated of all the rosin oils are the "light oils," or "rosin essence," and the work has been chiefly done by Pelletier and Walter. These chemists purified the rosin essence of commerce with sulphuric acid and caustic potash lye, and then fractionally distilled the product they obtained with careful observations of the boiling points. By this method they prepared the following constituents of light rosin oil in a pure state :

Rosin-naphtha or retin-naphtha.—This is a colourless mobile liquid darkened neither by air nor by light. It has a pleasant smell, a high refractive index, and will dissolve easily in alcohol, but not in water. It also dissolves in ether and in essential and fatty oils. It has a spicy burning taste and a specific gravity of 0·86. It boils at about 108° C. It is an isomer of toluol having the formula $C_7 H_8$.

Retinyl (*rosin oil*).—This is obtained in the fractional distillation of resin essence when the temperature has risen to 150° C., the boiling point of the liquid. It is colourless, but less mobile than resin naphtha. It has a specific gravity of 0·89, a bitter acrid taste, and is an isomer of cumol ($C_9 H_{12}$).

Retinol (*rosin train oil*).—This comes over at 240° and has a specific gravity of 0.9. It is a clear colourless liquid darkened by light. It has neither smell nor taste and is remarkable for its great solvent power for sulphur. Its formula is $C_8 H_8$.

Resisterene (*rosin-tallow*). —This consists of the distillation products of resin which come over between 250° and 350°. At ordinary temperature it is a solid crystalline substance, fusing at 67° and boiling at 325°. It is soluble in ether, absolute alcohol, essential oils, and liquid hydrocarbons.

As in rectifying rosin oils on a large scale lime or other alkalies are much used to free the products from acid, various chemists have investigated the behaviour of rosin when it is distilled, not by itself but mixed with lime. In one investigation by Fremy the rosin was mixed with eight times its weight of lime, this proportion being chosen to discover the action of the alkali upon the rosin. In practice the amount of lime used is of course much smaller, so that less of the products to be now mentioned will be obtained in the distillate.

In distilling rosin with lime, we get, together with small quantities of several other bodies, mainly two substances, which are of technical interest :

Resinon and Resineon.—Resinou is a very mobile colourless liquid with a sharp burning taste. It is very inflammable and boils at 78° C. Resineon boils at 148°, is less mobile than resinon, has a less pungent taste, and is less soluble in spirit. It is certain that quantities of these two bodies are always present in rosin oils which have been rectified with lime.

Later researches by Renard and Berthelot on the constitution of rosin oil led them to infer that it is a mixture of three hydrocarbons only. They say that the portion boiling below 346° consists of two of these, diterebenthylene (C_{20} H_{28}) and didecene (C_{20} H_{36}) The third hydrocarbon is called diterebenthyl and forms the largest part of the oil. Its formula is C_{20} H_{30}, it boils between 343° and 346°, has a specific gravity of 0·9088, and is an oily colourless liquid. According to Renard, the average composition of rosin oil is :—

Diterebenthyl - - - -	80	per cent.
Diterebenthylene - - -	10	,,
Didecene - - - - -	10	..

VIII. THE RECTIFICATION OF ROSIN OIL.

If the distillation is so conducted that all the products are collected in the same receiver a separation afterwards by rectification of the light and heavy oils is indispensable. If the distilling apparatus could be arranged for fractional distillation this separation could be effected from the first. But with the very large quantities of rosin that are treated at a time, this is scarcely possible, so that the fractionation must be done practically by means of a subsequent rectification.

This rectification can, however, be greatly simplified by collecting the first products of the crude distillates in two portions separately Several lots of each of these will give enough liquid for one rectification. If the final distillates are also collected in two separate portions according to the temperatures at which they came over, the rectification can be carried out in a very complete and simple manner. In many factories the rectification is still carried out in a very primitive way. The foreman judges of the progress of the distillation simply by the appearance of the distillate without giving any attention either to its boiling point or to its specific gravity. It is easy to see that a sharp separation of light, medium and heavy oils cannot be secured in such a manner. Both boiling point and specific gravity must be attended to in accurate fractionation and hence a thermometer must be put into the still and a hydrometer into the receiver. In working up the first portions of the crude distillate, which consist for the most part of the light oils, we observe that the thermometer after rising to a certain point remains stationary for a time, during which the hydrometer shows a constant specific gravity of the distillate. Presently however, the stream from the worm slackens and finally stops altogether. This shows that all the light oils have come over and that heavy oils only remain in the still. They are emptied out and distilled with the second portion of the crude distillate.

If the light oils contain water they froth very much in rectification and a still may boil over if only two-thirds full to begin with. The remedy is, of course, to get rid of the water as far as possible before rectifying. This is done by collecting large quantities of the first crude distillate and allowing them to stand before rectification until all the oils have risen to the surface and perfectly clear water can be thrown out from below them. The water drawn off will be opalescent as long as the separation is not complete.

In distilling the light oils it is advisable to keep the receiver unchanged while the distillate shows a specific gravity below 0.891 or 0.895 at the most. When the hydrometer shows that this specific gravity has been reached the receiver is changed, because the next portions of distillate belong to the heavy oils. The small quantities of light oils that have not yet come over are recovered in the rectification of the heavy oils.

Heavy oils are collected while the specific gravity is rising from 0.895 to 0.935. Between these limits we get what may be called "heavy blue oils." The temperature concerned is about 280 degrees C, and the distillate has a full wine yellow colour, is opalescent and shows a blue fluorescence at the edges. The heavy oils with a greater specific gravity than 0.935 are brownish and opalescent, and have a greenish fluorescence at the edges. Their average specific gravity is 0.940 and they are known by the name of brown rosin oil or train oil.

If a still heated by superheated steam is used the following are the average yields of the various products in percentage of the raw oil :—

Raw light rosin oil	11.25–12.
Raw heavy blue rosin oil	49 –50.5
Raw brown rosin oil	10.25–10.65
Pitch	18 –19
Water and acetic acid	5.5 –5.8

The Purification of Rosin Oil.

The foregoing remarks are concerned with the separation from one another of the three chief products of the distillation without freeing them from the small quantities of other bodies which they contain, and which although small in quantity cling to them with great pertinacity. The further purification from these is a chemical and not a distillation process. They consist chiefly of free acids together with small quantities of esters of the acids, so that alkalis are used for the purification, such as quicklime or caustic soda. Some processes less employed in actual practice are based on the use of powerful oxidising agents, such as nitric acid or a mixture of peroxide of lead and hydrochloric acid. Of late sulphuric acid has been extensively utilised in purifying the heavy oils.

Of the two alkalis, lime has the advantage in respect of price, but caustic soda in respect of convenience, for the lime in removing the impurities forms with many of them difficultly soluble compounds which make the cleaning of the apparatus troublesome. Caustic soda on the other hand makes (with all the impurities) soluble compounds which can be rinsed out without difficulty. Besides, the progress of chemical industries has now made caustic soda so cheap that for our purpose its excess of price over quicklime is really of little importance as compared with the superior advantages which it confers. In purifying light rosin oils, it is a greater point to get a colourless product, for such only are suitable for illuminating purposes.

THE PURIFICATION OF PINOLINE OR CAMPHENE.

Pinoline or camphene is the commercial name of light rosin oil which has been purified till it is colourless. When petroleum had not attained its present pre-eminence as a fuel for lamps, pinoline was a far more important substance than it is now, for it was the only liquid which would give a pure white light without smoke. Still even now, as it is always formed in the distillation of rosin, some use must be found for it, and this is only possible when it has been brought to a high state of purity.

To purify pinoline with caustic soda about 2 lb. of the solid hydrate are required for every 100 lbs. of the oil and are used dissolved in about two and a half gallons of water. The oil is put into a vessel provided with a stirrer and when that has been set going the soda lye is run in in a thin stream. As soon as all the lye is in the stirrer is stopped and the mixture is put into the still and quickly heated till the first drops of distillate appear. The heat is then let down so that the distillate comes over in only a thin stream. It often happens that at the beginning of the distillation a little liquid of a pale yellow colour comes over. This is repeated towards the end. These first and last distillates are collected separately to be repurified later. The intermediate portions are colourless, have an agreeable smell, and consist of pure pinoline. The purity of pinoline is generally judged by its behaviour when shaken up in a bottle. The air should not thicken it or make it brown. It must also burn in a good petroleum lamp with a bright flame and without smoke and leave no residue on the wick.

Pinolino was formerly purified with caustic lime and carbonate of soda, which form caustic soda and carbonate of lime during the process. The latter is insoluble and gave great trouble in cleaning out the still. The old method was to put into the still 100 gallons of water containing 20 lbs. of carbonate of soda in solution. A hundred pounds of quicklime were then put in 100 gallons of the crude light oil. The boiling attending the distillation mixed the liquids. It is easy to see that this process gives more trouble than the use of caustic soda, and in large factories it is almost entirely abandoned.

The Purification of the Heavy Rosin Oils.

The care that has to be bestowed upon the purification of the heavy oils depends upon the use for which they are intended. The purification consists in deodorisation, in the removal of free acid, and in the destruction of the substances which cause the fluorescence. A first class product must be of a pale yellow colour and have neither fluorescence nor smell. Such is the degree of purification required for a lubricant, for it is only such an oil that is without action on the metals with which it comes into contact in machinery.

Much heavy oil, however, is used for cart grease. Here the presence of fluorescence and smell are not of great importance, so that the purification can be confined to the removal of acid. The same applies to heavy rosin oils intended for the manufacture of lampblack or of printer's ink.

All the purification processes then include a destruction or neutralisation of any free acid present, combined if necessary with the removal of the fluorescent and of the odorous impurities. The chemicals used can be classed into two groups, mineral acids and caustic alkalis. Proper use of either will secure the result desired, but if caustic alkalis are used the bye products accompanying the purified oil are of value, whereby their use is to be preferred to that of mineral acid on financial grounds. We shall proceed to describe the most important of the many methods which have been devised, with the prefatory remark that many new processes have been boomed as important improvements when the changes were simply made for the sake of novelty.

The Herrburger Purification Method.

This process although rather expensive gives a perfectly colourless oil quite free from smell or fluorescence, and hence as regards the character of the product, answers every requirement. To get this highly purified oil we must confine ourselves to the second, third and fourth fifths of the rosin distillate and as the purification is the easier the less the distillate has been exposed to the air, it should follow the distillation as closely as may be. If this cannot be done the oil must be kept in full air tight vessels during the interval.

The oil to be purified is heated by a steam coil in a wooden vat provided with a stirrer. With continual stirring the temperature is gradually and steadily raised to 43 degrees C. When this point is reached 5 lbs. of nitric acid of 1.2 specific gravity to every 100 lbs. of oil are run in in a thin stream. When all the acid has been stirred in the rouser is stopped and the mass is left quiet for five to six hours. During this time the acid oxidises the impurities which form a tarry sediment. To ensure complete settlement the oil must be kept fluid by maintaining a temperature of about 40 degrees C. during the precipitation. The sediment is then withdrawn by a cock at the bottom of the vat. The oil remaining in the vat is washed again and again, each time by stirring it up in about a fifth of its bulk of hot water, until after standing the water can be run off quite colourless. The separated oil is then rectified with 2 per cent. of its weight of caustic soda lye of 25 degrees B. which removes the last traces of acid. On distilling this mixture of washed oil and caustic soda we get first a dark yellow distillate having a powerful odour. This is collected separately and mixed with some crude oil for another rectification. There follows a pure distillate and at the end of the distillation we get a little more yellow oil. The smell of this last fraction is, however, very slight and can be removed by exposure to the air for a few days. To get the greatest possible amount of colourless distillate, the distillation must proceed very slowly and the condensation must be thoroughly efficient. Unless these two conditions are fulfilled none of the distillate is ever really colourless.

The characteristic feature of Herrburger's process is the use of nitric acid. This powerful oxidising agent nitrifies the phenolic substances in the rosin oil and forms bodies soluble in water, especially in hot water so that the above described rinsing of the oil gets rid of them entirely. The tarry sediment consists of the products of the action of the acid upon other impurities whose nature is as yet but imperfectly understood.

The Kramer and Flammer Purification Method.

This process consists of a treatment with concentrated caustic soda lye followed by passing a strong current of air for a long time through the heated oil. In this way very large quantities of oxygen are brought into contact with the oil at a high temperature and exert upon it an action similar to that produced by the nitric acid in the Herrburger process and convert a number of the impurities into tar-like bodies.

According to the patent specification the process is carried out as follows:

The yellow oil is put into a vat which contains a steam coil opening freely into the oil. Here the oil is quickly brought to the boil by steam at about three atmospheres pressure. As soon as the boiling temperature is reached, 6 lbs. of caustic soda lye of 36 degrees B. are run in for every 100 lbs. of oil, and the boiling is continued until a sample of the contents of the vat separates quickly into an oily and watery layer on standing. This shows that the alkali has completely saponified the resinic acids originally present. When this saponification is incomplete a sample remains turbid and dark brown for a very long time.

The contents are now left to stand till the oil has formed a clear layer above the solution of rosin-soap. The latter is then run off and replaced by water having a volume equal to one quarter of that of the oil. Boiling is then resumed and kept up till a sample separates quickly on cooling. The water is then run off and mixed with the soap solution previously withdrawn for further manipulation.

The oil is now transferred to a vat with a false bottom and a steam coil. Here the oil is heated as quickly as possible to 60 or 80 degrees C. and a blast of air is blown through it in a number of fine streams, by means of an injector. After the blow has lasted for three hours, the temperature is raised to 100 degrees C. and the blast is made more powerful. The resulting oxidation further raises the temperature to 110 or even 115 degrees C. and the end of the process is indicated when the temperature ceases to rise.

This process gives an oil quite free from acid and from fluorescence. The smell is not entirely removed, but becomes very slight.

The rosin soap solution obtained as above described can be mixed with other soap in the course of manufacture, or can be used for making resinic acid by the following process:

A stream of carbonic acid is passed through the solution as long as any of it is absorbed. The solution soon becomes turbid with resinic acid liberated by the carbon dioxide from soap. The precipitated clots of resinic acid are collected and distilled with the next lot of rosin, while the solution from which they are removed and which consists simply of carbonate of soda, can be used for the preparation of caustic soda by means of quicklime.

The W. Kelbe Purification Method.

This process differs from that of Kramer and Flammer by no aeration of the oil following the treatment with caustic soda lye, Kelbe uses 16 litres of caustic soda lye of specific gravity 1·115 for every 100 kilogrammes of raw oil and heats to a temperature of 120 degrees C. This quantity of lye cannot, however, be regarded as an invariable quantity. Some crude oils will need more others less. The chief aim of the process, as in the case with all those in which caustic soda is used, is the saponification of the resinic acids in the crude oil, and as their amount varies, the amount of lye required wfll vary in proportion.

There are certain practical tests for ascertaining whether sufficient caustic soda lye has been used or not. If a sample is dark brown after long boiling and quickly separates into two layers on standing, a clear pale oil-layer above and a dark brown soap solution below, exactly the right amount of lye has been used. If solid matter is seen floating in the brown soap solution there is too little water present to dissolve all the soap. If some hot water is added the solid flakes will disappear.

According to Kelbe it is of special importance to keep the temperature strictly between certain limits. During the time subsequent to the half hour's boiling of the oil with the lye, the temperature must not fall below 50°, or rise above 65°. Below 50° the oil is thick, and is therefore unable to rise fast enough above the soap solution. If the temperature exceeds 65° the dark coloured soap solution imparts some of its colouring matter to the oil.

After repeated treatment with water the oil is put into large flat pans in which it is kept at a temperature of 60° to 80° C for a few days. Water and volatile oils then escape from it by evaporation. At the same time the fluorescent bodies are wholly destroyed, and the major portion of the odorous bodies also disappear, leaving a pale yellow oil almost entirely destitute of smell. It is easy to see that the lengthy heating of the oil in shallow vessels evaporates some substances and oxidises others, just as the air blast does in the Kramer-Flammer process.

The Hoffman Purification Method.

This purification process is based upon the fact that fuming sulphuric acid destroys the impurities in the crude oil. Hoffman gives the following directions:—Heat 200lb of the oil in a pan till it is of a thin consistency. Then remove the pan from the fire and drop into it, with continual stirring, 3lb of fuming sulphuric acid. Continue stirring for half an hour after the acid is all in, and then allow the mass to stand for twelve hours. At the expiration of this time the clear oil is ladled off from the sediment and washed with hot water till the water ceases to acquire an acid reaction to blue litmus paper. This treatment with fuming sulphuric acid gives an oil of a rather dark yellow, and having a slight smell, but no fluorescence. To get the oil pale and entirely without odour a second treatment must be resorted to which may be regarded as an oxidation and bleaching process.

Every 100lb of oil is shaken with 1lb of water in a glass flask, and then treated with 8lb of hydrochloric acid previously diluted with one gallon of water. The acid is gradually stirred into the oil. There are next added 1lb of red lead and 5lb more of the dilute acid. The whole mass is repeatedly shaken up in the flask during several days. The oil is then decanted from the precipitate and washed with water till the wash water has no acid reaction. The oil is finally exposed for a week in white glass flasks to the full rays of the sun. It is said that the final production is a very pale yellow oil, which answers excellently as a lubricant when mixed with half its volume of a fatty oil.

The various processes for rosin oil purification just described certainly all produce a non-fluorescent acid-free oil with very little smell, but the cost of carrying them out varies considerably. That of Hoffman is the most expensive of all, for it requires a considerable quantity of chemicals, and even then a special bleaching of the oil by light is required to get it pale enough.

It is hardly necessary to say that the best process for purifying rosin oil is not one that gives worthless bye-products, or such as are only available as fuel at the best, but one which enables the impurities removed from the oil to be used.

According to our experience the best method is that in which the purification is effected with caustic soda lye, for the considerable quantity of resinic acid occurring in all crude oils forms rosin soap with the soda. The soap can be recovered either by mere evaporation of the solutions and can be used by the resin factory, or sold to soap boilers for admixture with common soaps. But as caustic soda alone will not give a product of sufficient purity the treatment with the alkali must be followed not only by a washing but by a further purification. As this is always done by oxidation, it can be effected either by nitric acid or with the aid of heat by a blast of air maintained for a sufficient time.

In whatever manner the oil has been purified it retains in most cases a pale yellow colour. To get rid of this, as well as of the last traces of smell, it is much to be recommended that the oil should be exposed to the sun in large glass flasks (the glass must be white) for a time varying with the strength of the sunlight. As a rule about two weeks' exposure is sufficient, and the final result is the purest product that can be obtained from rosin-distillates. The flasks must be quite full and stoppered air-tight.

Distillation of Rosin with Alkalies.

If rosin is dry distilled by itself, we always get together with the rosin oils and gases considerable quantities of acid, which come over with the oil and dissolve in it, and must be separated from it by one of the previously described methods.

There is, however, a special method of distillation which prevents any acid from coming over with the distillate and keeps it back in the still. This method consists in putting alkali into the still with the rosin; all the acids then combine with the alkali as they are formed and converted into soap-like products. For this purpose two alkalies are available, caustic lime (quicklime) and caustic soda. The former has the advantage of being the cheaper of the two, the latter of forming soluble salts of immediate usability with the resinic acid. If quicklime is used, the only way of getting the acid from the lime soap is to boil it with dilute sulphuric acid, when the lime is precipitated as sulphate, and the liberated resinic acid collects in clots which swim in the liquid. These clots can be distilled with a charge of rosin. It is much simpler to distil with caustic soda. A rather strong solution of the alkali in water is made and put with the rosin into the still. The distillate is then free from acid from the beginning, and the rosin-soda-soap solution can be removed as it is, and be either concentrated by further evaporation or be used as an addition to an ordinary soap. This production of rosin soap in the still necessarily diminishes the yield of crude oil, but that obtained is far easier to purify, and the rosin soap remaining in the still is a very valuable product. Calculation has shown that rosin distillation with caustic soda is more economical than distillation without. It must not be forgotten that caustic soda must always be used in the purification of the crude oil, and that the quantity of the refined oil is distinctly smaller than that of the crude distillate.

The amount of quicklime or caustic soda to be put in a given quantity of rosin depends upon the origin and the age of the latter. As a rule, however, from 6 to 9 per cent. of the weight of the rosin is sufficient of either.

The Manufacture of Cod Oil.

The oil known as cod oil, or train oil, is made in a special way, somewhat resembling the distillation of rosin with caustic soda. The name of cod oil is used because the oil has a consistency resembling that of genuine cod liver oil, so much so that it is used for the adulteration of the latter.

Cod oil is made from the blue oils of rosin distillation in the following manner :—

The blue oils are boiled with water for twenty-four hours, the water being renewed as it evaporates. The water is then run off and the necessary amount of caustic soda lye of 37° B is added to the oil. The resulting saponification of the free resinic acid produces a buttery, nearly solid mass, which is distilled as long as it will give a liquid distillate. The seconds cod oil thus made is put in iron vessels, having on the bottom a layer of powdered plaster of Paris, the result is first quality cod oil. The plaster frees the oil from water.

Cod oils are rather yellow in colour, but are almost always the palest of all the rosin oils. They ought to show nothing more than a very faint bluish fluorescence. Cod oil differs from the other kinds of rosin oil also in its strikingly high boiling point. While the boiling point of the ordinary heavy oils is in the neighbourhood of 200° C., cod oils boil at 230 to 240°. Most of the samples we have tested boiled at 235°.

The General Properties of Rosin Oils.

The products which are put upon the market as rosin oils show considerable differences in properties according to the extent to which they have been refined. They are liquids, and being more or less thick, flow in a manner recalling very old linseed oil. Their colour is usually yellow or orange yellow, but is sometimes a pure red.

If we look slantingly at the surface of rosin oil in a glass we nearly always see a blue fluorescence, although it may be only very slight. The smell of a properly prepared rosin oil is pleasant and its specific gravity must not be below 0·96 or above 0·99, and the boiling point must not vary much from 200° C. When strongly cooled, rosin oil does not change greatly in consistency, and this property is very favourable to its use as a lubricant, for the change in temperature between summer and winter makes very little difference in the consistency of a rosin oil lubricant.

Very highly purified rosin oil has no trace of any acid reaction and is unaltered by exposure to the air, *i.e.*, does not resinify. By treatment with twice its volume of strong alcohol, rosin oil is dissolved, usually to the extent of from 50 to 70 per cent. The properties mentioned above in this paragraph apply, however, to very carefully purified oils. Ordinary commercial oil has generally a fairly strong acid reaction, and easily absorbs considerable amounts of oxygen when exposed in thin layers to the air. By so doing it thickens sometimes so much that it will scarcely flow. Hence rosin oil in its action to the air greatly resembles ordinary drying oils, and for that reason can replace them to some extent in varnish making. On account of its affinity for oxygen, and the thickening it undergoes by absorbing it, rosin oil cannot be used by itself as a lubricant, as the resinification of the oil would make it very difficult to clean the machinery. The researches of Bach show the great absorptive power of rosin oil for oxygen. According to him, one gramme of ordinary rosin oil can absorb no less than 181 c.c. of oxygen, and a gramme of cod oil of specific gravity 0·963 absorbs 76·3 c.c. of oxygen. Compared with these numbers the amount of oxygen absorbed by petroleum is very small, being only 0·45 to 0·7 c.c. per gramme of oil.

Nowadays rosin oil has practically gone out as a lubricant for machinery, having been replaced by the heavy petroleum oils which answer the purpose admirably. The low price of petroleum has also put products on the market as rosin oil, which contain only 60 to 70 per cent. of that substance, the rest consisting of heavy petroleum oils. Unfortunately, too, rosin oil is a very favourite adulterant of fatty oils, especially linseed oil and rape oil, and the finer kinds of seal and fish oil are also falsified with rosin oil. It has already been mentioned that some rosin oils are known as train oil from their outward resemblance to fish oil.

The chief use of rosin oil is unquestionably in the manufacture of cart grease, for making special kinds of paint, and finally for the production of the finest kind of lampblack, such as is used for preparing art printing inks. Other uses of rosin oil are for making brewers' pitch, and for making a low quality of sealing wax for bottles.

Rosin Oil Varnishes.

The great absorptive power for oxygen possessed by rosin oil, whereby it thickens greatly, has produced attempts to substitute it for drying fatty oils such as linseed in the manufacture of varnishes. Great advantage is to be secured by so doing, as rosin oil is very much cheaper than linseed oil. The method adopted is analogous to that followed with linseed oil. As is well known linseed oil is made drying by heating it for a considerable time with certain substances. These were formerly exclusively lead compounds, especially litharge and red lead, and occasionally sugar of lead. Experience showed, however, that although the products obtained left nothing to be desired in respect of rapid drying, they had the drawback of darkening very much in colour when exposed to the air. This effect is produced by the sulphuretted hydrogen in the atmosphere, which gradually converts the lead in the oil into black sulphide of lead.

To prevent this darkening of the oil of varnishes made with it, other driers are now employed which are not so affected. The chief two of these new bodies are manganous borate and linoleate.

According to various published recipes a good resin oil varnish can be made by dissolving melted rosin in hot rosin oil, adding driers (manganous borate for choice) and continuing the heat till the proper consistency is obtained. Some makers say also that a very good varnish is got by adding certain sulphur compounds to the above recipe.

A recipe containing no sulphur, the object of which certainly seems doubtful from a chemical point of view, is as follows :—

Colophony from pine-rosin	100 lb.
Thin rosin oil	60 ,,
Boiled linseed oil	40 ,,
Pinoline	100 ,,
Oil of turpentine	80 ,,

The ingredients are mixed as follows :—The boiled oil is made rather hot, and the colophony, first broken into small pieces, is dissolved in it by stirring. The temperature is then allowed to fall a little, the rosin oil is next added, then the pinoline, and finally the turpentine. After thorough mixing, the whole is allowed to stand and settle, and is lastly filtered through a cloth. As regards the pinoline, it seems advisable to add it only when no other use can be found for it, and if it is cheaper than oil of turpentine. If these conditions do not hold it is best to replace the pinoline by as much turpentine as may be necessary. Pinoline is inflammable and very volatile, and its use is accompanied by considerable risk of fire. It may easily happen in making a varnish with it by artificial light that the liquid may be accidentally ignited. The varnish may even catch fire spontaneously in stock.

It is true that the use of the pine-resin colophony makes the varnish much cheaper, but it has also its shady side. Experience has proved that rosin oil varnishes containing such considerable quantities of pine-resin colophony as mentioned in the recipe soon lose their beautiful lustre, and when they are exposed alternately to wet and to the rays of the sun, as when the varnish is used for garden seats, wooden fencing, etc., soon turn dull and whitish. To get a varnish free from these drawbacks it is therefore advisable to take much smaller quantities of the colophony than directed. If instead of 100lb., 25 to 50lb. are taken, a varnish is obtained which has the above disadvantage in a less marked degree.

To prepare rosin oil varnishes, which are really comparable to fatty oil varnishes in appearance and durability, it is urgently to be recommended that the pine colophony should be replaced entirely or almost entirely by a resinate. As we have already seen the resinates have the property of forming very thick solutions. which, on the evaporation of the solvent, form hard lustrous coats offering very great resistance to the weather. The best resinates to use are those of alumina and zinc.

Still cheaper rosin oil varnishes can be made by partly converting the pine-resin colophony into resinate of lime, and using this instead of the pure colophony. The pine-resin colophony is melted and heated till it begins to emit heavy fumes, and quicklime in very fine powder is then stirred into it. The lime readily unites with the resinic acid of the colophony. Care must be taken that no excess of lime is left in the mass. When resinate of lime is used the varnish is prepared as follows :—The rosin oil and linseed oil are first heated together and then the resinate of lime is carefully stirred in until a sample of the mass is rather thick after quick cooling. The oil of turpentine is then added, and if the varnish is not then thin enough rosin oil is added in small portions with heating until the cold varnish has the proper consistency.

IX. THE MANUFACTURE OF PATENT LUBRICANTS.

The names patent lubricant and patent cart grease are applied to some special compositions which have now nearly driven every other lubricant for ordinary vehicles out of the market. Their chief superiority lies in their cheapness, as they cost much less than lubricants containing ordinary fats or soaps. They have, however, another valuable property not possessed by fatty lubricants. They are practically unaffected by changes of temperature. It is often necessary to use at least two kinds of fat lubricant, one in the winter and the other in the summer. The patent lubricants can be used unchanged from one end of the year to the other.

The fundamental ingredient of every patent lubricant is what may be called a lime-rosin-oil soap. This we shall call patent soap for short.

If an attempt is made to saponify rosin oil with caustic soda or potash, it will be found that even the strongest lyes have very little action on it. At the outside 1¼ per cent. of the oil is saponified. If this soap is decomposed with an acid, semi-solid brown bodies having a strong resinous smell are formed, which greatly resemble the acids of the rosin from which the oil was made.

Rosin oil behaves very differently with the alkaline earths. On boiling rosin oil with quicklime the alkali gradually dissolves, and forms a mass of about the consistency of butter. This is dark brown in colour and can be mixed with more rosin oil, so as to possess any desired degree of thickness. This mass is the patent soap. Chemically, however, it does not show the properties of a soap. It is probably a compound of lime with hydrocarbons, but this has not been investigated.

In making the patent soap on a large scale, the first consideration is to procure the right sort of lime. Many limestones contain carbonate of magnesia, and the lime burnt from them consequently contains magnesia. Such lime is quite unsuitable for patent soap making. The lime used must be pure. This is easily tested by slaking it. Pure lime rapidly absorbs large quantities of water, soon breaks up, evolves much heat and forms a very soft powder. If there is much magnesia in the lime it slakes on the other hand very slowly, and evolves but little heat. Pure slaked lime is called fat lime, because when mixed with water it can be spread like butter, a property of which magnesiferous lime is entirely destitute, so that it is called thin.

The first process, a pure lime having been procured, is to slake it. The lime is put upon a floor free from sand and dust, and a weight of water equal to half that of the lime is provided and scattered. over the lime from a rose, and this gradual sprinkling is continued uniformly over all the lime till it has completely fallen into a fine dry powder. Care must be taken not to use more water than is absolutely necessary to produce this result, as the slaked lime must show no visible sign of the presence of water.

As even the very best limes may contain grains of sand the powder must be sifted through a fine sieve to separate them. Slaked lime greedily absorbs carbonic acid from the air. To prevent it from doing so it must be casked as soon as made and sifted, and closed up so that the air cannot get at it.

To make the patent soap, a large iron pan is required, provided with a stirrer. The rosin oil is put into this and heated nearly to boiling over an open fire. Then add 80lbs. of slaked lime for every 100 of oil, stirring it in a little at a time, waiting after each addition till the white streaks at first formed disappear. After the lime is all in, the heating and stirring are still kept up till the contents of the pan form a tough syrupy mass, which can be drawn out into long threads by dipping in a spatula. The process must not be regarded as complete before a sample of the mass painted thinly on glass shows a perfectly uniform yellow or brownish appearance, without the least trace of white specks of undissolved lime. Such specks must not appear even under a magnifying glass.

Tho finished patent soap will keep fit for use for an indefinite time, but as it must be heated to make a lubricant from it, it is a saving of fuel to work it up as soon as made and while it is still hot, so that it has simply to be ladled into another pan.

An enormous number of recipes for making patent cart greases are available, and they are usually distinguished by their colour, as, for instance, yellow, blue, green, brown, and black greases. The compositions are also put upon the market under various names. Whatever kind of cart grease is to be made from the patent soap, the process must be carried out in special shallow pans with vertical sides and slightly convex bottom, so that the pan is shallowest in the middle. The pan contains a stirrer with two vertical blades which come close to the bottom, and also close to the sides. The blades are perforated with a large number of circular holes. The pans are put into an oven in such a way that their bottoms are heated all over, but their sides are only partially exposed to the fire.

The colour of the finished grease depends partly on that of the rosin oil, but partly upon colouring matters added simply as colour. Fillings are also added in many cases to cheapen the product or to increase its consistency, and are inactive bodies having no lubricating power of their own.

Manufacture of Patent Blue Cart Grease.

To make this favourite grease, gas tar and petroleum are used as well as rosin oil. Both of them are first worked up with lime, and then mixed with blue rosin oil and the patent soap. The process may be described as follows :—

Put into the pan just described 50 lbs. of tar oil, and 50 lbs. of blue petroleum. Heat and stir in gradually 150 lbs. of slaked lime. Make a uniform mixture by continual stirring, as the patent soap must not be added before. About 225 lbs. of it are then heated and stirred in in portions, stirring the whole time. There is above the pan a cask containing 200 lbs of heavy blue resin oil, which can be run into the pan through a pipe with a cock, having a rose at the end. The cock is opened until the oil falls into the mixture from the rose into the pan in large drops or very thin streams. The stirring is kept up all the time the blue oil is flowing into the pan. As it does so the contents of the pan become thicker and thicker, and it is a question solely of practical experience to be able to judge when the proportions between the various ingredients are correct. The mass must be made so thick that it will not drop from a stick which has been plunged into it, but stick on in a tough lump. When this point is reached, the behaviour of a sample is observed at the ordinary temperature in a small dish of thin sheet iron. Some of the hot mass is spread out in the dish, the outside of which is then cooled in cold water. If the cold mass has the consistency of soft butter and spreads like it, the process is finished. If, however, the sample appears too solid, like cheese or hard butter, it has been badly made and must be rectified by adding more of some ingredient. The one usually required is the patent soap, and it should be first tried by adding to the entire mass two or three per cent. of it. If this does not put matters right, blue resin oil must be added with prolonged stirring. When the lubricant has been got right, the stirrer is stopped, and the mass is scooped out with shallow ladles into the drums in which it is sold. A new lot can then be started in the mixing pan at once.

Experience has amply demonstrated that rosin oils of various origin behave very differently in the manufacture of cart grease, and even rosin oil from the same distillation will act different'y according to whether it is fresh or stale. In general, old oil which has been long in stock is better than the freshly distilled product, which almost always gives lubricants which are too soft. If it is found, in spite of every care in the making, that the product is not stiff enough, a little more of the patent soap will put matters right. One per cent of it will often be sufficient to impart the proper buttery consistency.

In large works, the safest plan is to mix the rosin oils from a large number of distillations in one big cistern. The result is a very great quantity of rosin oil of one quality, and the directions given above will enable anyone without much difficulty to make very large quantities of unexceptionable patent cart grease.

Patent Coloured Cart Grease.

The blue patent cart grease owes its colour to being made with strongly fluorescent rosin oil. Greases of any other colour must be artificially dyed. Yellow grease, for example, is dyed with a decoction of curcuma root in soda lye. It is best to prepare the decoction separately, and to put into the mixing pan, at the very beginning of the manufacture, the amount which experience has shown will be required for colouring the batch. No definite directions as to the quantity can be given, and every maker has his own recipe. It is advisable for every one to have a recipe arrived at by trial, recording the amounts of ingredients used in getting a good result, and to keep to that recipe thereafter, so that the goods are always turned out exactly the same. Experience has shown that buyers go chiefly by the colour, a property which has no influence on the lubricating powers of the stuff, and therefore care must be taken that successive makes of product are always exactly identical in this respect. We now proceed to give instructions for making the yellow and black greases.

Patent Yellow Cart Grease.

This is usually made from a mixture of the two patent soaps, I. and II., which are put together with boiled heavy rosin oil into the mixing pan, and gradually mixed with soda lye which has been well boiled with curcuma root broken small. The dye in the root dissolves in the lye, which is then freed from the remainder of the pieces of root by being run through a sieve. The following is an approved recipe for making yellow cart grease :

Patent soap I.	30 lbs.
Patent soap II.	60 ,,
Boiled heavy rosin oil	500 ,,
Caustic soda lye of 16°B	16 ,,
Curcuma root	2 ,,

Patent Black Cart Grease.

There are two different ways of making this product, either by colouring an ordinary patent cart grease with lampblack, or by using black wood-tar pitch for the manufacture.

In the first case the lampblack must be perfectly mixed with a portion of the patent soap, and then mixed thoroughly with the rest. Put into the mixing pan 2 lbs. of lampblack, and pour over it 30 lbs. of patent soap I. Heat and stir until no separate particles of lampblack can be seen. Then stir in 15 lbs. of patent soap II., and finally 250 lbs. of heavy blue boiled rosin oil. The mass is kept warm, and stirred until perfectly uniform. The best test of the completion of the mixing is the appearance of a very thin coat on glass held up to the light. A properly made mass will then, by transmitted light, appear of a perfectly uniform dark blackish-brown colour.

To make black patent cart grease with black pitch the pitch left in rosin distillation, which still contains a distinct amount of rosin, is employed. First heat together in the mixing pan 25 lbs. of patent soap I. and 30 lbs. of patent soap II. Then heat 50 lbs. of pitch in a separate vessel and pour it, together with 200 lbs. of heavy rosin oil, into the soaps in the mixing pan, stirring well the whole time to prevent the pitch from going into lumps, and to mix it in uniformly.

Making Cart Grease with Soluble Rosin Soap.

Patent cart grease can be made without lime and then the advantage is secured that some of the rosin oil can be replaced by petroleum, which is cheaper. The first of this class of processes was proposed by Dr. Herrburger, and has been retained to the present day in all its essential features, although other makers have made large and different changes in the details. The following is a process which gives good results:—Put into a pan 350 lbs. of residue from petroleum refining, such as a crude dark oil, or vaseline in its most impure form, or the thick heavy oil obtained by distilling coal tar. When one of these substances has been made sufficiently fluid by heat, stir in 120 lbs. to 130 lbs. of crude rosin oil, and finally 212 lbs. to 234 lbs. of rosin soap in powder. As soon as the soda soap is in, stir immediately, as the mass begins to froth very much, and keep stirring until froth has all disappeared. Then stir in from 50 lbs. to 60 lbs. of melted rosin, according to the consistency required, and continue stirring until the cooling mass has become too stiff to be stirred.

The rosin soap used must be perfectly dry. The best way to dry it is to let it dry in small pieces in the air as much as it will, and then to cut it into shavings, which are then thoroughly dried by artificial heat. Then powder the soap and sift the powder.

Filling of Patent Cart Greases.

To make the greases as cheap as possible, the device is resorted to of filling them up to 50 per cent. of suitable mineral substances. It is, however, a mistake to add to the grease as much as half its weight of such bodies. As a rule the percentage should not exceed 20, and in all cases the filling must be perfectly mixed with the lubricant. The chief substances used are levigated clay, powdered barytes, and soap stone. The barytes is, however, not to be recommended, and that for two reasons. The first is its great specific gravity, so that even if 20 per cent. of it is added, the bulk of the stuff is not much increased. The second is that the powder is crystalline, and is rather hard, so that however finely it may be ground, the hard particles, so far from lessening friction, increase it materially. Powdered soapstone is a crypto-crystalline mineral of very low hardness, and having a greasy feel, it is, therefore, a better filling for cart grease than levigated clay, although the latter is cheaper. The filling must be stirred into the cart grease while it is still hot and soft. The best method is to shake the powdered filling through a sieve on to the hot grease, stirring the latter continuously all the time.

X. ROSIN SOAPS OR RESINATES.

If an animal or vegetable fat is boiled with caustic soda or potash, the fat is decomposed into glycerine, which mixes with the water present, while the fatty acids combine with the alkali to form salts, which are freely soluble in water and are known in ordinary language as soaps.

Ordinary rosin contains a body having the properties of an acid, so that it dissolves on boiling with the caustic alkali, giving a solution of an alkaline resinate. As these compounds of resinic acid with alkalies behave in exactly the same way as the corresponding compounds of fatty acids, we may call them soaps by analogy, and they are known as rosin soaps or resinates. Just as among the salts of the fatty acids, only those with soda or potash are soluble in water, so the compounds of resinic acid are soluble only when the alkali is potash or soda. The resinates of metals other than potassium or sodium are quite insoluble. If therefore we pour into the solution of a soluble fatty acid or resinic acid soap, one of another salt, a precipitate is formed of the insoluble soap. If for example we add a solution of sulphate of zinc, we get a precipitate of zinc resinate, while the potash or soda remains in solution as sulphate. The insoluble fatty acid soaps are of no value, except for the preparation of pure fatty acid, but the case is very different with the rosin soaps. The insoluble rosin soaps are extensively used for the preparation of excellent lakes, varnishes and paints. They constitute also the only material which enables us to prepare enamel colours for the decoration of porcelain and stoneware.

Nevertheless the use of the insoluble rosin soaps has been somewhat hampered in industry by the want of recipes which will enable the manufacturer to prepare them in a suitable manner.

Manufacture of Ordinary Rosin Soap.

Resinate of soda, which is ordinary rosin soap, is the starting point in the manufacture of the insoluble rosin soaps, As pure resin soaps, that is, soaps made from rosin and caustic soda and without any fat, are not upon the market, the manufacturer must make his own, which he can do in the following manner,

It would be possible to proceed by boiling the rosin to be saponified in a solution of carbonate of soda. The acids of the rosin expel the carbonic acid from the soda with effervescence, and themselves combine with the alkali, whereby the soap is produced. The disadvantage of this method, however, is the very long boiling necessary to saponify all the rosin, and moreover the rosin clots together in the hot liquid, and as the clots present only a small surface in proportion to their size, they offer considerable resistance to the action.

It is, therefore, better to begin with caustic soda, and to finish the saponification with carbonate of soda after a certain stage has been reached. If we boil even a weak solution of caustic soda with rosin, it is soon observable that the liquid becomes milky, and an examination of a few drops of it with a magnifying glass will show that the rosin is spread about in it, in the form of very small drops. Thus the first effect of the action of the caustic lye is to form an emulsion, for an emulsion consists in an enormous number of minute drops of one liquid suspended in the mass of another. Thus, so far there is no saponification, and it is only after long boiling that the rosin dissolves and the liquid becomes clear. If carbonate of soda is added, as soon as the emulsion is formed the immense surface then offered by the rosin causes a vigorous action, accompanied by rapid evolution of carbonic acid. The process is carried out in a large iron pot heated over an open fire, or better by steam.

FIG. 25.

A pan for heating on an open fire is shown in Fig. 25 together with the fire-place.

In Fig. 26 is shown a plan of the setting of the pan; the arrows show the direction in which the gases from the fire travel. In working on a large scale, however, it is decidedly better to use a pan heated by steam.

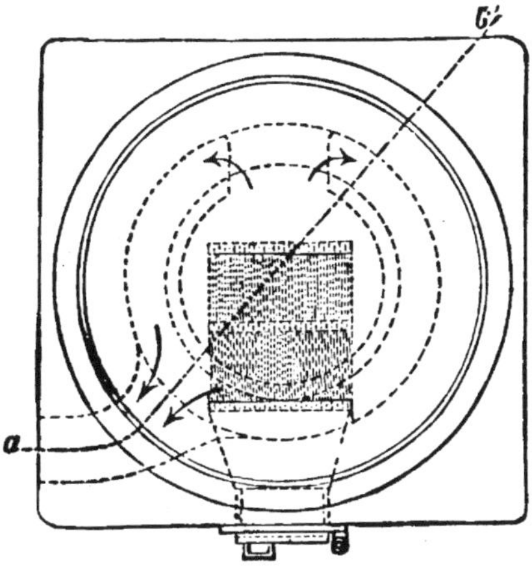

FIG. 26.

FIG. 27.

A steam-heated pan for a small factory is shown in Fig. 27. The steam enters the jacket through the hollow bearing on the left, and the tap seen below the jacket serves to drain out the condensed water. The pan is emptied by tilting it with the endless screw and bevel wheel.

A larger pan for dealing with larger quantities at a time is shown in Fig. 28. As the figure shows, it contains a stirrer composed of rods and chains, which is driven from beneath the pan.

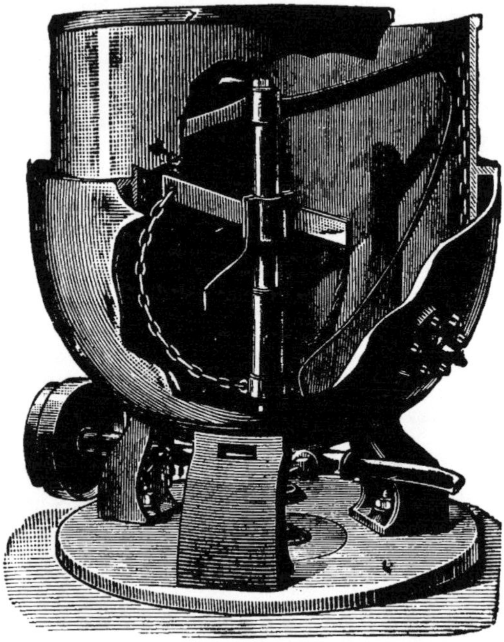

FIG. 28.

The work is begun by putting weak caustic soda lye into the pan and setting the stirrer in motion. The rosin, first broken up rather small, is then allowed to fall into the pan in a thin stream. As soon as enough rosin for one operation has been added, the pan is boiled till a complete emulsion has been formed. Some of the solution of carbonate of soda is then added, and the boiling is continued till the effervescense has nearly ceased. More soda is added and the boiling is continued until the effervescence again stops, and so on, until a sample of the liquid is yellowish and rather thick, like a weak solution of glue. If the liquid is not stirred a small quantity of melted rosin often settles to the bottom of the pan ; if this happens it shows that enough soda has not been used to complete the saponification. It is however advisable to leave a little rosin unsaponified, for only by so doing can we make sure that there is no excess of soda in the liquid. The residue of the rosin can be used up in the next batch.

As soon as saponification has taken place the steam is shut off and the stirrer is stopped. The liquid in the pan is allowed to stand a few hours, in order that the mechanical impurities that were in the resin, such as small stones and bits of wood, may sink to the bottom, where they are generally found embedded in the unsaponified rosin. The soap solution is poured into the wooden vat in which it is kept through a large closely woven cloth to filter it. The vats are covered with a wooden cover to keep out dust.

When concentrated solution of carbonate of soda has been used, the soap solution when cool is rather thick. It is not diluted until it is required for use.

Pure rosin soap has a yellowish or yellowish brown colour, in some cases a dark brown, the colour depending upon the kind of rosin used. Pale American rosin gives a fine pale yellow soap.

Resin soap belongs to the soft soaps, and at ordinary temperatures it has the consistency of soft butter. It dissolves easily in water, and this when even slightly warm takes up large quantities of it. It can be used for washing just as it is, but as soft soaps are cheaper than hard ones, most soap-boilers employ rosin soap only as an addition to ordinary soda soaps which are very much harder. By using the very much cheaper rosin soap in this way they make a very large profit.

Soap-boilers usually work up various mixtures of tallow, palm oil, and rosin. Good results are obtained, for example, with 600lbs. to 700lbs. of tallow, 50lbs. to 100lbs. of palm oil and 200lbs. to 300lbs. of rosin. The best procedure is to saponify the tallow first by itself, then to add the palm oil and saponify it. The rosin is saponified by itself in a separate vessel, and the finished rosin soap is added to the mixture of the other two. The mixture is then boiled and crutched till it is clear

If it is desired to make insoluble rosin soap, it is obvious that no fat can be used; neither is it necessary to concentrate by boiling. Care must be taken that the soda rosin soap used contains no excess of soda, or it will be unfit for use.

The Insoluble Rosin Soaps.

All insoluble rosin soaps are made by one and the same method. A certain quantity of the rosin soap solution is diluted with water, and a salt of the metal of which the resinate is required is prepared in solution and added to the soap solution. The desired resinate is then obtained. As it is impossible to take the two substances exactly in the proportion which leaves no excess of either, a test must be made to ascertain which of them is in excess after the precipitation is complete. For this purpose some of the liquid is filtered through paper into a glass, and mixed with a little of the metallic solution. The appearance of a precipitate shows that more metallic salts must be added to complete the precipitation. A little is therefore added, and after careful stirring the test is repeated. When the filtered sample ceases to give a precipitate with the metallic salt it is certain that all the rosin soap has been decomposed. As there may be an excess of metal in the soluble form, another sample must be filtered and tested with a solution of rosin soap, which will give a precipitate in the presence of an excess of metal. These tests are continued alternately, adding each time what the last test has shown to be necessary, until no precipitate is given, either by the metallic salt or by rosin soap.

The work may be much shortened by determining the specific gravity of the soap solution as well as that of the metallic salt solution, and in all subsequent operations using solutions of the same specific gravity. The repeated testing of samples will then only be necessary with the first lot of insoluble soap manufactured. Care must be taken to add together all the amounts of rosin soap solution used and all the amounts of metallic salt solution used, for the main mass. In subsequent operations all we have to do is to use these quantities to be sure that neither substance is in excess. In any case one or two tests at the most will be sufficient to get the proportions exactly right.

Different insoluble metallic resinates have of course different properties. The resinates of lime, magnesia, and alumina are white and when wet are remarkable for being extremely bulky, but become much less so when thoroughly dry. In the dry state these three resinates form a very light powder. They dissolve readily in essential oils and remain behind in colourless masses when a thin coat of solution is allowed to evaporate into the air. The resinates of the heavy metals show characteristic colours in cases where the oxides of the metals form coloured salts. Ferrous resinate is usually green like ferrous salts in general, and, like them, turns brown by oxidation on exposure to the air. Nickel and copper resinates are green, while those of lead, zinc, and bismuth are white.

In the manufacture of resin lakes and varnishes, as well as paints, the white resinates are most important, as their solutions give colourless masses on drying. The resinates of alumina and zinc are speciallysuitable on account of their low price, and we shall describe the production of these two resinates somewhat more in detail.

Resinate of Alumina.

This resinate is always made from sulphate of alumina, which can be bought very pure and very cheaply. It is largely used by dyers and calico printers as a mordant, and is met with on the market in the form of brick-shaped pieces readily soluble in water. The solution is usually turbid from the presence of small quantities of an insoluble basic salt. There are two ways of getting the solution clear. One is to let it stand and decant it from the sediment formed and the other is to convert the basic salt into the soluble salt. This is done by gradually adding sulphuric acid, stirring after each addition, and taking care to add the acid sufficiently cautiously to avoid any excess. An excess added after the solution has become clear will set free resinic acid which will contaminate the resinate afterwards made. The formula of sulphate of alumina is $Al_2O_3(SO_4)_3 + H_2O$ and it contains 15 per cent. of alumina. The following table shows the percentage of pure anhydrous sulphate of alumina and also of commercial sulphate of alumina in solutions having the stated specific gravities at 15 degrees C. :—

Pure anhyorous sulphate of alumina.	Commercial sulphate of alumina.	Percentage.
1·0170	1·0069	1
1·0270	1·0141	2
1·0370	1·0221	3
1·0470	1·0299	4
1·0569	1·0377	5
1·0670	1·6416	6
1·0768	1·0481	7
1·0870	1·0592	8
1·0968	1·0650	9
1·0171	1·0730	10
1·1171	1·0794	11
1·1270	1·0860	12
1·1369	1·0960	13
1·1467	1·1059	14
1·1574	1·1097	15
1·1668	1·1169	16
1·1770	1·1199	17
1·1876	1·1269	18
1·1971	1·1339	19
1·2074	1·1440	20
1·2168	1·1448	21
1·2274	1·1589	22
1·2375	1·1628	23
1·2473	1·1689	24
1·2572	1·1798	25

In making the resinate large vats are used provided with spigots arranged one above another in a vertical line, and four inches apart. A saturated solution of the sulphate of alumina is made in a small vessel, poured into the big one, and there diluted with water till the density shows that the desired concentration has been reached. It is made as concentrated as is possible without preventing the mass from being stirrable after the soap has been added. Over the vat containing the alumina solution another is arranged with a draw-off cock near its bottom. This is filled with a concentrated solution of rosin soap. When everything is ready the liquid in the large vat is stirred vigorously and the rosin soap is run in from the upper vessel. A very voluminous precipitate is immediately formed consisting of resinate of alumina, and the stirring is continued while the soap is running in. When it may be supposed that enough soap has been used, a sample is filtered from the lower vat and tested as above described. It must give a precipitate neither with a solution of sulphate of alumina nor with a solution of resin soap. When this point has been reached, the stirring is stopped and the liquid allowed to stand. The bulky precipitate settles only slowly, and the supernatant liquid clears from above downwards. The liquid is run off as it clears by opening the spigots one after another, beginning at the top. The final residue at the bottom of the vat is a white paste somewhat resembling thick starch.

To separate the resinate of alumina from the liquid it contains, smaller vats with their bottoms full of tiny holes are used. Sacks are fitted accurately into these vats so that the closely woven linen of which they are made lies close against the side. At the upper edge of these sacks a strong string is sewn in for lifting the sacks out of the vats. The paste at the bottom of the precipitating vat is ladled into the sacks till they are quite full, and the liquid present runs out through the holes in the bottom of the vat. When the mass in the sack has shrunk by reason of this, more of the paste should be put into it. This is continued till the precipitate in the sack comes up to within four inches of the edge of the vat. When the precipitate is completely drained it is washed by pouring clean water into the sack and allowing it to drain again. Three washings with water are done. The resinate of alumina is then ready for drying. If there is plenty of time and the weather is warm, the resinate can be dried by hanging it up in the sacks. The drying can be accelerated by first squeezing the sacks in a press so as to drive out as much water as possible. Great care must be taken not to burst the sack which is easily done if the pressure is too strong or too suddenly applied. It is possible to get the resinate of alumina so nearly dry by pressing, that it can be spread out on strips of sacking to dry.

Even when it appears completely dried in the air there is still water adhering to the mass, and this must be artificially removed before the resinate of alumina can be dissolved. The simplest method of completing the drying is to spread the resinate on strips of sacking arranged one above another in a drying room. The temperature of this room is kept at about 40 degrees C. for a few hours and the dried resinate is then filled while still warm into tightly closed receptacles, where it remains till it is to be used. This complete drying is of great importance, for otherwise the resinate will not give clear solutions, nor will the solutions give coats of the proper lustre when dry.

Pure dry resinate of alumina is a white friable mass of small specific gravity. It gives a characteristic dry feeling to the fingers. It dissolves in various solvents, but most easily in essential oils, in which it first swells up into a jelly and then quickly dissolves. In making varnishes and paint, however, the solvent generally used is rectified oil of turpentine. When the turpentine evaporates on exposure to the air, a colourless, fairly hard mass is left which offers great resistance to atmospheric influences and is therefore a very good substitute for an oil varnish.

RESINATE OF ZINC.

For making this resinate the cheapest zinc salt on the market is used. This is zinc sulphate or zinc vitriol. Commercial sulphate of zinc usually contains iron and would not give a colourless lake but one made yellow by the presence of iron. It is, however, easy to purify the zinc sulphate in the following manner :—

Add carefully to a solution of the salt a small quantity of caustic soda lye until a white precipitate begins to appear. Then stir the liquid thoroughly. The precipitate will then disappear because the zinc sulphate often contains much free acid which dissolves the precipitate of hydrated oxide of zinc produced by the caustic soda. It is necessary, therefore, to add the soda until the precipitate is permanent. In a few days an ochre-yellow sediment has settled in the bottom, for the oxide of iron in the solution is thrown down by the zinc oxide, which itself redissolves. In this way the zinc salt can be entirely freed from iron and will give a perfectly white resin soap.

Zinc resinate is prepared in an exactly similar manner to that which has just been described for resinate of alumina. In this case also the two solutions are used of known specific gravity, so that a single trial will show how much of each solution to take in every subsequent operation. The following table shows the connection between the percentage of zinc sulphate in the water and the specific gravity of the solution :—

Specific gravity of solution.	Percentage of zinc sulphate.
1·028	5
1·059	10
1·090	15
1·123	20
1·157	25
1·193	30
1·231	35
1·270	40
1·310	45
1·353	50
1·398	55
1·445	60

As the specific gravity of zinc resinate is much greater than that of resinate of alumina the precipitate settles more rapidly to the bottom of the precipitating vat, and the washing of the soap can be done sooner. Pure zinc resinate appears when dried in rather heavy white pieces which are very easily reduced to a fine powder.

Resinate can be made with soluble salts of any heavy metal and a solution of resinate of soda. In this way the resinates of copper lead, silver and tin are prepared, but in varnish making as well as in the preparation of paint little use is made of any but the resinates of alumina and zinc.

A special use is made of the resinates of bismuth, iron, and of that of alumina to some extent in ceramics to produce various effects on porcelain or pottery. These resinates are usually known as enamels or glazes and will be described in a later section of the book.

XI. THE MANUFACTURE OF RESINATE VARNISHES.

Resinate varnishes are prepared in a simple manner, for it is sufficient to dissolve the resinate in an appropriate solvent. The best, because it dissolves the resinates most easily, is rectified oil of turpentine. Their solubility in turpentine is so great that it is possible to make solutions of every degree of consistency from that of water to that of a fairly thick syrup.

Solutions of resinate of alumina are pale to dark yellow according to the concentration, while those of zinc resinates are rather brown.

As it is necessary to use perfectly clear solutions, it appears most convenient to dissolve the resinates in large glass vessels in which the solutions are kept till they have become perfectly clear. The resinate is put into the flask, which must be quite dry, in pieces about the size of a hazel nut until the flask is about one-third full. Enough turpentine is then put in to quite cover the pieces and the flask is corked and allowed to stand for a few hours. At first the resinate swells considerably and passes into a gelatinous mass. The action is accelerated by frequently shaking the flask and when the gelatinisation is complete, more turpentine is added, a little at a time until we have a solution of the desired consistency, when it can be used as it is for a varnish or can be mixed with pigments to make a paint. It is necessary in order not to make the solution too thin to add the turpentine with great care. As soon as the proper quantity of turpentine is in, the solution is shaken up once thoroughly and the flask is put in a place where it cannot be shaken so that the solution may clarify. It may then be decanted quite clear from the sediment at the bottom.

The preliminary gelatinisation of the resinate makes solution at the ordinary temperature rather a slow process and care must be taken to keep the jelly as loose as possible by constantly shaking the flask. As this is rather troublesome in working on a large scale it is advisable to somewhat modify the process, and to effect the solution by the aid of heat. This is best done in the following manner :—

The resinate to be dissolved is put into a large iron pot well tinned inside. This pot is put into another and the space is filled with water. This water is boiled over an ordinary fire, after as much turpentine has been poured over the resinate as will completely cover it. The resinate and turpentine are stirred with a broad paddle until they have reached the temperature of the boiling water. The resinate runs down as a thick mass of the appearance and consistency of honey from the paddle, and when this is the case the additional turpentine is added until the thoroughly stirred mass runs through the paddle in a continuous stream. We then begin to take small samples, which are painted on to a sheet of glass so as to rapidly cool to the ordinary temperature. The cold mass rapidly assumes an appearance much resembling that of set glue. When it does so it is a sign that more turpentine must be added, and this must be continued until a cold sample has the consistency of good boiled oil. This shows that the proper amount of turpentine has been added. The solution in the pot is then allowed to cool and poured out into large glass flasks in which it can stand to clarify. The sediment of undissolved particles clings with some tenacity to the bottom of the flask so that it is very easy to decant the pure liquid.

The amount of turpentine oil required to make a solution of any given consistency varies greatly with the nature of the resinate. Zinc resinate dissolves much more easily than alumina resinate Twenty-two gallons of oil of turpentine are sufficient to give a thin. solution with a hundredweight of zinc resinate, while the same quantity of alumina resinate requires three times as much turpentine and even then forms no proper solution but a thick buttery mass. The latter is, therefore, the resinate to be preferred on economic grounds both for the preparation of varnishes and of mixed paints.

To obtain fine lustrous and durable varnishes from solutions of resinates it is advisable to use the solution rather thin, and to apply a second coat as soon as the first is dry. The resistance of these resinate varnishes to atmospheric influences is very great, and long exposure to the continuous action of air, sun and rain leaves the lustre of the varnish unaltered. These varnishes also offer great resistance to the action of chemicals. Frequent washing of a varnished object with soap and water affects the lustre only very slightly and the coats resist solutions of carbonate of soda or potash even better than they do soap ; but fairly concentrated solutions of caustic soda or potash attack the varnish rather sharply, by decomposing the resinates with separation of alumina or zinc oxide while the alkali combines with the resinic acid and forms a soap soluble in water.

It is safe to assert, however, that as regards resistance to atmospheric influences and the action of chemicals the resinate varnishes are quite equal to the best fatty oil varnishes, while they have the advantage of being very much cheaper.

RESINATE PAINTS.

The solutions of resinates in oil of turpentine prepared by one of the previously described processes can be mixed with various powdered pigments to produce bodies which can be used for paint after a simple addition of a suitable amount of turpentine.

The paints are made by pouring over the perfectly dry fine powder of the pigment enough solution of resinate to form a paste, thin enough to be worked up in an ordinary colour mill. The first passage through the mill usually makes the mass somewhat stiffer so that a little more resinate solution must be added to it. The mass is returned to the mill often enough to produce a perfectly uniform paint and the number of times this has to be done depends upon the fineness of the powered pigment. The best way of finding when the proper point has been reached, is to take a small sample and paint it thinly on glass. If no separate pigment particles can be seen and if the colour is perfectly uniform the process may be regarded as complete. The sample coat must lose its turpentine rapidly by evaporation and show when quite dry a smooth and lustrous surface.

There is another simple method of preparing resinate paints by which it is possible to obtain a large amount of saleable goods in a short time. The pigment used is reduced to fine powder as well as the resinate. The powders must be very carefully dried, for even a very small amount of moisture has a bad effect upon the properties of the paint, especially upon its lustre. The two powders must be most thoroughly mixed in the proper proportions. This is best done by putting them into a rotating cask which is slowly turned until the mixture is complete. The time required for this can only be determined by experiment, so that from time to time the rotation of the cask is stopped and a sample of the powder is taken out and rubbed up with a spatula with a little oil of turpentine. If the mixture is right it will form a thick paste even with a very small quantity of turpentine and will with a little more turpentine form a mass of the consistency of a thick oil colour and will work well with the brush. A sample painted on a board must rapidly dry to a lustrous coat and cover the wood completely. When the sample shows that the proper point has been arrived at the powder is taken out of the cask and thinned with turpentine till it will go through the colour mill. As the particles of resinate are uniformly distributed through the whole mass, they are rapidly dissolved by the turpentine and a paint fit for use can generally be obtained after two passages through the colour mill.

It is advisable to use no more turpentine than is necessary to make a very stiff paste, for the latter can always be diluted at will before use. If enough turpentine is added at first to make the mass of the proper consistency with which to paint the larger part of the solvent will separate out in the drums in which the paint is stocked. We then find in the lower part of the drum a thick mass which it is very difficult to mix again with the liquid over it. This appearance is especially common when the pigment mixed with the resinate has a high specific gravity as in the case, for example, with all lead pigments.

In painting with resinate paints it may often be noticed that a short time after drying the coats become traversed with a large number of minute cracks and flake off in places. This is a sign that too much resinate in proportion to the pigment has been taken. On the evaporation of the oil of turpentine the resinate shrinks considerably, and thereby cracks the layer of paint. It is therefore necessary to make preliminary experiments with every pigment in order to decide in what proportion it must be mixed with the resinate in order to get a paint which will not crack on drying. Full security, however, against the cracking of the paint is afforded if, in addition to the oil of turpentine, a very small amount of rosin oil or of drying linseed oil prepared in the cold with borate of manganese is used, for these bodies remain liquid after the evaporation of the turpentine, drying later by resinification. The paint certainly dries somewhat more slowly, but it retains a certain amount of elasticity so that no cracking occurs on cooling. Hence no cracking takes place either in the paint or in the varnish, and both retain their lustre completely.

The quantities of rosin oil or boiled oil to be used for this purpose are always very small, and, as a rule, an amount equivalent to 5 per cent. of the turpentine oil used is sufficient. The rosin oil or boiled oil is added at the same time as the oil of turpentine, in order that all the liquid may be uniformly distributed through the entire paint. As finished paints soon thicken in the air from the evaporation of the turpentine they must be enclosed at once in air-tight drums.

Paintings executed with these paints have a smooth shining appearance, and in this respect surpass ordinary paints made with pigments and boiled oil. To secure the durability of the lustre, it is a good plan as soon as the resinate paints are perfectly dry to give an extra coat of dilute varnish. As the mass in which the particles of pigment are embedded in a wrapping of pure resinate and that which results from the drying of the extra coat with pure varnish are entirely identical, they adhere firmly together and no cracking takes place either in the paint or in the varnish over it. In the case of paintings executed with boiled oil paints and varnished over with a fatty oil varnish fine hair cracks invariably appear in time no matter how fine a varnish has been used. Hence the varnish loses its lustre. These hair cracks result from unequal expansion, as changes of temperature act differently on the paint and on the varnish. As we have already remarked the mass containing the pigment is in the case of resinate paints the same as that constituting the varnish, so that both expand equally on heating and contract equally, and no cracking occurs.

ENAMELS.

The preparations known by this name in ceramics consist of the resinates of various heavy metals prepared by the method already described. Some enamels, however, have nothing in common with these but the name, and usually consist of solutions of metallic salts from which the metal is easily set free, so that by their means very beautiful and extremely thin layers of metal can be produced upon porcelain, stone ware or glass.

We proceed to describe briefly the preparation of the commoner enamels.

ALUMINA ENAMEL.

This kind of enamel produces when applied to a white ground of porcelain or any finely glazed earthenware an optical effect considerably resembling that produced by mother-of-pearl. This enamel is prepared by dissolving ordinary resinate of alumina in oil of turpentine or other essential oil and painting the solution upon the earthenware with a brush. It must be remembered that the effect of the enamel varies very greatly with the concentration of the solution used, as solutions of different strength give on the earthenware coats of different thickness. Hence to get any given effect we must determine by previous experiments the particular concentration of solution required. When the solvent has evaporated the resinate can be fixed by a gentle firing.

BISMUTH ENAMEL.

This enamel is used in porcelain painting as a flux for the purpose of fixing other enamels or to dilute them and make their colour paler. Used by itself it produces hardly any effect,

Bismuth enamel is prepared by dissolving bismuth in concentrated nitric acid under a chimney with a good draft, which will carry away the large quantity of unwholesome fumes produced. When the metal is all dissolved, the solution is allowed to stand for a time to deposit any insoluble residue. The clear solution is then decanted into clean water. The water should have four or five hundred times the bulk of the bismuth solution. Contact between the two liquids forms basic nitrate of bismuth, which goes down as a dazzling white precipitate and gradually sinks to the bottom of the water. When all the precipitate is perfectly settled, the water is carefully decanted away from the very dilute solution of neutral nitrate of bismuth with several lots of water. The precipitate is then dried and is known as bismuth white.

The recipe for making flux from the bismuth white is as follows:

Bismuth white - - - - - - -	10 lbs.
Rosin - - - - - - - -	30 ,,
Oil of turpentine or lavender - - - - -	75 ,,

The rosin is put into a large porcelain basin and melted on a sand bath. The bismuth white is then added to it a little at a time, and thoroughly amalgamated with it by careful stirring. The mass toughens and becomes brown. It is then diluted with a little of the oil. More bismuth white is then stirred in, and so on, and when the mixture is complete the whole mass is stirred until cold.

Lead Enamel.

For second rate porcelain goods and ordinary pottery, lead enamel is used instead of bismuth enamel, being much cheaper. Lead enamel is made in an exactly similar manner to bismuth enamel, using nitrate of lead instead of the basic nitrate of bismuth.

Bismuth lustre is often used to lighten the shade of coloured enamels.

For example, iron enamel used alone gives an orange red which can be made pale at will by mixing it with flux. The same is true for all other coloured enamels.

Iron Enamel.

This enamel is made in various ways. In one method a solution of chloride of iron in water is added to a solution of rosin soap until the formation of a precipitate ceases. The precipitate is then filtered off, washed with water and dried. Dried iron enamel is a reddish brown solid. The second process is to melt together equal weights of rosin and turpentine and then to stir in as much nitrate of iron as there is rosin. The mass is stirred with constant heating until it is perfectly uniform. Used by itself iron enamel gives an orange red, which can be toned down to any extent by an admixture of flux.

Uranium Enamel.

This enamel is made like iron enamel except that nitrate of uranium is used instead of nitrate of iron. By itself uranium enamel gives a warm nankeen yellow, but when diluted with flux it gives all shades down to a pale straw yellow.

Chromium Enamel.

Make a solution of bichromate of potash in water, add to it a little sulphuric acid, warm it and add to it spirit of wine, a little at a time. The solution froths up strongly and its beautiful red colour changes to violet. When the solution is cold it is precipitated with a solution of rosin soap. The precipitate is washed and dried on porous plates. This enamel gives a beautiful green.

Cadmium Enamel.

This is made by precipitating a solution of rosin soap with one of sulphate of cadmium. Used alone it gives a beautiful mother-of-pearl lustre, and mixed with enamels of other colours it imparts to them the same quailty.

Cobalt Enamel.

This is made like iron enamel, substituting nitrate of cobalt for the nitrate of iron.

Metal Enamels.

The so-called metal enamels differ from those just described in consisting of very thin layers of pure metal, which on the opaque ground show their full lustre and their characteristic colour. The chief of them are gold enamel and platinum enamel, both of which are used to a fairly large extent in ceramics.

Platinum Enamel.

Thin platinum foil is cut up into very small pieces and put into a mixture of nitric acid with one third of its bulk of hydrochloric acid. This liquid is called *aqua regia* and dissolves the platinum slowly. The solution is evaporated at a gentle heat in a porcelain dish, but without boiling it, until a solid reddish brown residue is left. This residue consists of tetrachloride of platinum and is dissolved in strong spirit. To produce platinum enamel on porcelain a little of this solution is mixed with oil of lavender and applied with a brush. When the porcelain has been fired the painted places appear covered with a shining white coating of metallic platinum.

Gold Enamel.

To prepare this enamel we use either thin sheet gold or gold wire. The metal is dissolved in *aqua regia* like the platinum. On careful evaporation a reddish brown mass remains, consisting of trichloride of gold. This is dissolved in spirit and the solution is painted on the places to be gilt. The thickness of the gilding depends upon the concentration of the gold solution used. The more concentrated the solution the thicker the gilding. Fairly thick gilding can be polished to a high degree of lustre with agate or polishing steel. The more dilute the spirituous solution is, the thinner is the gilding, but it always forms upon an opaque surface a completely continuous lustrous coating. With extremely dilute solutions the layer of gold is so thin that it transmits a green light, but it still shows its characteristic yellow lustre by reflected light. Gilding so thin as that just described is of course very easily rubbed off leaving no trace behind, while the thicker coats which have been polished are so durable that even when the object has been in use for years the gilding remains entirely unaltered.

Resinate Pigments Proper.

The resinate paints described in a previous section, consist as we have said of various powdered pigments mixed with a solution of resinate of alumina or zinc. On evaporation of the oil of turpentine the lustrous coat of resinate in which the coloured powder is embedded remains behind.

There are, however, resinate pigments proper which differ from those already described in being composed of metallic resinates and basic coal tar dyes. They possess all the beautiful hues belonging to those dyes.

The basic coal tar dyes are a collection of dyes frequently called aniline dyes, which have the property of combining with acids. It is very easy to find whether a commercial coal tar dye belongs to the basic or to the acid group. All that has to be done is to add a little solution of tannin to a solution of the dye in a test tube. If the tannin produces a precipitate the dye is a basic one and can be used for the preparation of resinate pigments. If on the other hand there is no precipitate the dye belongs to the acid group and is useless for our purpose.

As examples of basic dyes available for making resinate pigments we may mention fuchsine, methyl-violet, brilliant-green, auramine, saffranine, etc. Even this limited selection shows what a great variety of resinate pigments can be prepared.

The resinate pigments are made by making rather a dilute solution of rosin soap in water and pouring into it an aqueous solution of a suitable basic coal tar dye. The intensity of the pigment produced depends upon the amount of the coal tar dye, so that in order to get a pigment with the same depth of colour every time the same weights of ingredients must always be used, taking as a starting point the weight of rosin contained in the soap employed. In other words we find by experience what colour we get when we use an amount of dye equivalent to 1, 2, 3, per cent., etc., of the rosin. Having found once for all the colour produced by each of these different proportions we are always in a position to re-produce it. When the rosin soap and the dye solution have been thoroughly mixed a solution of the metallic salt is poured into them in a thin stream. The salt generally used is sulphate of magnesia, alumina, or zinc. As the insoluble rosin soap is formed it carries down the dye with it. This carrying down of the dye necessarily makes the liquid paler, and the addition of the metallic salt is continued until it becomes entirely colourless. A small excess of metallic salt does no harm. As soon as the precipitation is complete, the liquid is allowed to stand in the precipitating vessel until it has cleared. The liquid is then decanted, the precipitate is washed repeatedly with clean water, and finally allowed to dry on a filter cloth in a place protected from dust. In working on a large scale it is advisable to do the washing and drying of the precipitate in a centrifugal machine. The drum of the machine must be provided with a closely woven linen cloth fastened to its sides. The vessel containing the precipitate is placed above the centrifugal which is then set in motion, but not too rapidly, while the precipitate is allowed to run into it. At first the liquid from the centrifugal is strongly coloured and it is collected and put back into the drum. As soon as the layer of pigment on the sides of the drum has attained a certain thickness, nothing passes through the linen but clear liquid. When all the precipitate is in the drum a thin stream of water is allowed to flow upon it. This water rapidly drives out the soluble salts mixed with the precipitate.

After a short time the water can be turned off and the drum is then set to rotate at its highest speed. The centrifugal force then drives out the water present so completely that the precipitate is converted from a paste into a fairly dry crumbly mass which is lifted out of the drum in the sack and completely dried at a gentle heat.

During the whole process of making these resinate pigments care must be taken not to expose them unnecessarily to light, as they have the remarkable property of gradually losing their solubility in the direct rays of the sun.

The resinate pigments dissolve in various solvents with considerable ease. Benzol, ether, alcohol or chloroform can be used as a solvent. If a solution of a resinate pigment is wanted, only thoroughly dried pigments and pure solvents, such as very strong alcohol or rectified benzol, must be used. If a clear solution prepared in this way is painted in a thin coat the evaporation of the solvent leaves behind a coloured but quite transparent sheet of the resinate pigment which shows a glassy lustre and a considerable degree of hardness.

In making the resinate paints, any metallic solution can be used which will give a white precipitate with rosin soap. Hence the most important salts are those of zinc, magnesia, lead and tin. Salts which produce colours of their own independently of the rosin, such as salts of iron or copper, are useless for our purpose, as the paints made with them would come out of a different colour from that of the tar colour which has to be fixed. As, of course, it is desirable to use the cheapest bodies possible, zinc or magnesia salts should always be used for resinate paints in which no special shade of the colour is wanted. The cheapest zinc salt on the market is the sulphate and among magnesium salts the chloride. Both these salts give resinates, of almost exactly the same colour, but the powders differ in specific gravity. Resinate paints made with magnesia are far lighter than those made with zinc. For the same reason too, zinc paints are more easily levigated than magnesia paints as they settle to the bottom of the water much faster. The use of lead should be avoided for making resinate paints. All lead paints become discoloured and finally black by exposure to the air and sometimes do so very quickly, from the formation of black sulphide of lead by sulphuretted hydrogen in the atmosphere, where it very commonly occurs in towns.

The same organic dye will give different colours with different resinates. For example, magenta gives with magnesia a paint showing the full beauty of the red of the dye, while if tin is used, the paint will be violet and any number of shades can be obtained by mixing the tin paint with the magnesia paint in various proportions.

Carbon Pigments.

XII. THE MANUFACTURE OF LAMPBLACK AND LAMPBLACK PIGMENTS.

There is really only one really black pigment known. All other so-called black bodies are either dark blue or dark brown, but amorphous carbon or lampblack is really black.

The theory of the conditions leading up to the liberation of carbon in an amorphous state from combination are very important to the manufacturer and must therefore be briefly stated. Every organic compound contains carbon, and can be burnt by heating it in the air, whereupon its carbon forms carbonic acid and its hydrogen forms water. If, however, there is not enough oxygen to effect this latter conversion, it is, as a rule, only the hydrogen that is completely burnt. Some of the carbon is set free in the solid form as a deep black powder, known as lampblack, and when impure as soot. Alcohol contains both carbon and hydrogen and oxygen as well. When burnt it is completely consumed to carbonic acid and water. Turpentine and petroleum are both hydrocarbons, but contain far more carbon in proportion to the hydrogen than alcohol does. If either of them is set fire to in an open dish it burns with a dull yellow flame from which a thick cloud of finely divided carbon arises as smoke. Such combustion is called imperfect. It is possible to burn either petroleum or turpentine completely so as to form nothing but carbonic acid and water. This is what happens in a properly constructed petroleum lamp. Here the wick takes the form of a hollow cylinder so that oxygen comes into contact with the inside as well as the outside of the flame. The chimney strengthens the draught of air through the flame and no soot escapes from the top of the chimney with the combustion gases. But the combination in these lamps takes place in two sharply marked phases which can be easily recognised by the appearance of the flame. Just above the wick is a blue, barely luminous flame which consists chiefly of burning hydrogen. The upper part of the flame is luminous and the better the lamp the whiter it is and the worse the lamp, *i.e.*, the less the draught, the redder and yellower it is. At first there is an imperfeet combustion of the petroleum which burns with liberation of carbon, but the particulars are carried up by the rising gases heated white hot and gradually consumed. This short description clearly indicates how any combustible must be burnt to get the greatest possible quantity of its carbon in the form of lampblack. The substance must be allowed to come into contact with just so much oxygen as will keep it alight, but no more, or unnecessary amounts of the carbon will be burnt to carbonic acid. In this case the flame will give more light and heat than before, and this must be carefully guarded against.

Any organic body rich in carbon can be used for lampblack making. The substances chiefly used have for a long time been rosin, turpentine and the resinous wood of various trees, especially the roots. Later animal and vegetable fats were added to the list, and of recent times large quantities of lampblack have been made from the bye-products of petroleum refining. It can also be made from coal tar pitch. In using all these only a small portion of their carbon is burnt. The rest is set free in a finely divided state with a large number of products of dry distillation clinging to them. The more there is of these bodies the less pure is the black of the carbon, and they also give it a greasy nature. The lampblack must therefore be subjected to a special purification process to destroy these bodies. Lampblack can also be made by a process which is quite different from imperfect combustion, and which must be regarded as a sort of dry distillation. If any organic body rich in carbon is enclosed in a vessel with only a narrow opening and it is heated, it decomposes, forming both liquid and gaseous products containing all the hydrogen and oxygen of the original substance. The excess of carbon remains behind as an amorphous coal, still containing the mineral substances originally present in the form of organic salts. Among vegetable bodies used in this way, wood must first be mentioned. Wine lees and grape stalks are also used. Among the animal substances employed for the purpose are bones, fragments of horn, and dried blood. To obtain pure lampblack from the resulting products, the mineral substances must be dissolved out from it. The solvents are then rinsed away with water, and on drying pure amorphous carbon is left which can be used in the same way as smoke black freed from *its* impurities.

MAKING LAMPBLACK FROM RESINOUS WOODS.

We put the description of this method of manufacture first because it is unquestionably the oldest of all. Rude and primitive though it is it is still practised in countries rich in pine trees. The roots of these trees have a yellow wood which is extremely rich in essential oil, hardly differing from oil of turpentine, and could easily be extracted by breaking the wood up small. But this rational method is not practised and the people content themselves with burning the roots in a limited supply of air, thereby procuring large quantities of impure black, a little wood tar, and an essential oil having a very disagreeable smell from contamination with em-pyreumatic bodies and called crude pine oil. A smoke hut, such as can still be seen in many places, is represented in Fig. 29.

FIG. 29.

It consists of a hut R, the size of an ordinary living room and made of rough boards. The ceiling of this chamber consists of a coarse fabric which hangs down in great folds about a quarter of the height of the room in depth. Strings are fixed to various parts of it which are all joined to one which passes over a pulley. On pulling the one string the cloth is lifted and falls again by its own weight when let go. The chamber is usually 16 or 17 feet square and about 10 feet high. It is covered with a light roof to keep the rain from the ceiling-cloth. A brick tunnel K, about a foot square opens into one side of R. This tunnel is 17 to 20 feet long and slopes gently. It communicates with the brickwork chamber F, which can be closed by an iron door provided with a damper. The chopped up roots are burnt on the sloping floor on the part near the tunnel. On the door side a narrow channel leads to a ditch G, which collects the tar and the empyreumatic oil.

The oven F, is charged with chopped wood which is set fire to. The iron door is then shut and the damper is kept open enough to maintain a thick red flame. The combustion gases carry the soot up the incline of K into R. Here they deposit most of the black on the floor and pass away through the ceiling. This in time gets so choked with soot that the draught is stopped, whereupon the string is worked to shake the soot down. Only the wood nearest the iron door really burns and hot combustion gases from it have to pass through the remainder of the wood which is hereby subjected to a sort of dry distillation, giving up their essential oil and a part of their tar which flow into the ditch G. It is of the highest importance to keep the flame in the oven just right. If it gets too strong from an excess of draught, the soot accumulated in the tunnel and the chamber R may be ignited and the result would almost certainly be that the whole place would be burnt to the ground. The soot in R being very finely divided and mixed with inflammable empyreumatic bodies is very easily set on fire and is at once ignited by contact with a burning substance. Hence the entry of the flame into the chamber may have disastrous consequences in addition to contaminating the lampblack with ashes. The lampblack must not be removed from the chamber R till the oven is quite cold, the fire having been extinguished by completely shutting the damper A narrow door in the side of R is then opened and the lampblack is taken out with a wide flat wooden peel. No trouble is taken to clear the black out very thoroughly as small scrapings are blown about and lost. The appearance of the black removed will at once show whether the draught has been properly managed. Laid upon white paper it must appear a pure black and streaks of it must not look brown except just at the ends. If the air has been deficient the black is distinctly brownish, is greasy and makes a brown mark if rubbed on the paper. In this case the temperature has been too low, and quantities of tarry matter have escaped decomposition. Although in this case the black is less valuable than the pure product its formation is not regarded as a fault by the soot burners who look at the fact that the yield is greater than when the pure substance is obtained. The brown product can be purified afterwards.

A large amount of the lampblack is used just as it leaves the soot chamber for all purposes where cheapness is the great desideratum, such as making grey washes with milk of lime, or very cheap paints with resin oil, or printer's ink which need not be absolutely black. All the products so made, however, have the disadvantage of slow drying an account of the tarry impurities in the lampblack. If a newspaper is printed with one of these cheap inks, it smudges when folded, "sets off" and dirties the hand. Attempts have been made to improve the primitive construction above described by making the sides like Venetian blinds. Such a modification is given in Fig. 30.

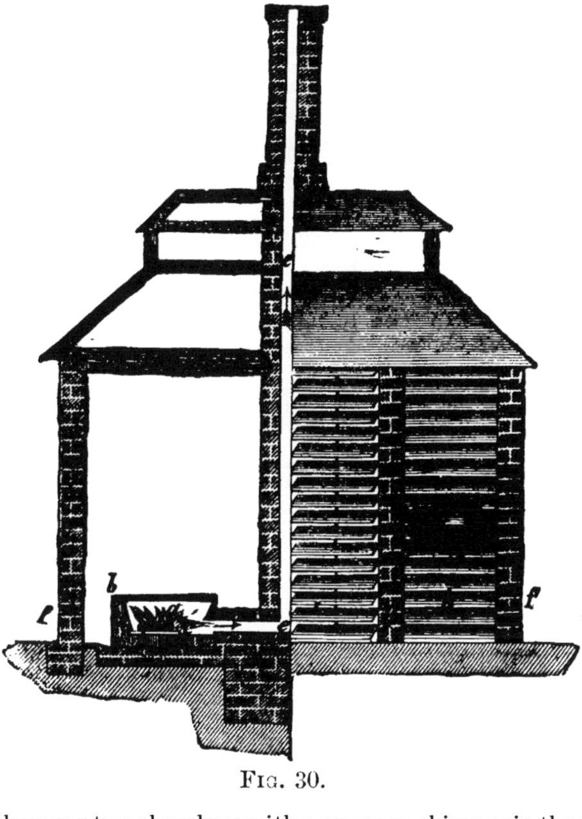

FIG. 30.

Here there are two chambers with a common chimney in the middle. The oven and other parts are constructed as above described. The slates of the sides are used to regulate the draught, but this cannot be properly done, and the contrivance loses more black than the old arrangement with a linen ceiling.

PURIFICATION OF LAMPBLACK.

To free the crude lampblack from tarry matter it is heated red hot whereby the tars are either decomposed with the formation of more lampblack or are volatilised. Simple as this operation may appear it must be executed with certain precautions, or some of the lampblack will be converted into shiny hard crusts instead of a dead black powder. These crusts must be ground and levigated before their carbon is fit for use.

The ignition, which is sometimes wrongly called calcination is best done in tubes of very thick iron plate and having their outsides protected from the fire by a plaster of loam, water and chopped straw. This coating must be renewed from time to time. The iron tube is closed air-tight at one end and has a lid at the other, having a small opening in the middle of it. The tubes used have a diameter of from 6 to 8 inches and length of from 4 to 5 feet. They are placed horizontally in an oven which is traversed by strong iron bars, such as old railway metal, so that several rows of tubes can be arranged at different heights. The tubes must be so placed that the openings in the lids are outside the furnace and can be observed. The crude black is shaken loosely into the tubes until they are full except the last two inches. The lid is then put on and the joint is carefully luted with loam, so that the tube is air-tight except for the hole in the lid. It is important that the black should be loose in the tubes, so that the gases developed by the heat may escape freely. When the oven is filled with tubes the fire is lighted. When the lower layer of tubes has been at a bright red heat for about two hours the fire is increased until the top row is red hot. The tubes are then taken out, the holes in the lid are stopped with loam and the tubes are left to get thoroughly cool. If the tubes are opened before they are cold the finely divided carbon will begin to glow the moment it comes into contact with the air by condensing the oxygen from it and will soon break into flame. Cold lampblack will not catch fire spontaneously on exposure to the air. To keep the lampblack from again becoming impure it should be packed up the moment it is taken out of the tubes. The best things to pack it in are tinplate canisters with well fitting lids and of such sizes that they hold one, two, or three kilograms of lampblack. When the canisters are filled and closed the lid is sealed by gumming on a strip of paper.

When the tubes begin to be red hot, a stream of gas will be found to be issuing from the small holes in the lids, which will burn with a bright flame. The gas increases in volume at first, but finally disappears. It is produced from the tarry bodies in the lampblack which are either volatilised or decomposed with liberation of solid carbon.

The going out of the flame is a sign that all the tarry matters are gone and hence that the operation is concluded, and the tubes can be removed from the fire. Properly purified lampblack when removed from the cold tubes must be perfectly black and without lustre. It must, when rubbed thin on a sheet give a dead black without greasing the paper. If the mark is brown the tubes have not been made hot enough to destroy all the impurities present.

In purifying the lampblack the temperature must be raised to a red heat gradually. If the heating is too quick, the sudden evolution of gas may blow off the lids of the tubes, whereupon the contents will burn. Even when this does not happen, the too rapid heating may cause the formation of the above-mentioned shiny crusts which require grinding and levigation to make the carbon serviceable. Properly ignited lampblack can be used in the manufacture even of the finest printing inks.

XIII. LAMPBLACK CHAMBERS.

It is easy to see that the description already given refers to a chamber of very primitive construction, which requires little art to convert it to one which is more suitable for the purpose. On watching one of these primitive arrangements in action it is easy to see that there is always a fairly thick black cloud hanging over the ceiling cloth. This consists of combustion gases and the finest of the carbon particles swept by the gases through the meshes of the cloth. Hence the best and finest particles are lost, and only the coarser ones are collected from the chamber, being too heavy for the gases to sweep away.

It is, however, possible with a proper arrangement to separate the lampblack so thoroughly from the combustion gases that only very small quantities are lost, and so that an automatic sorting of the carbon in coarser and finer particles takes place. The essential point in such an erection is to make the combustion gases travel a considerable distance before they can escape. The cooling and friction which they then undergo so lessens their speed that they deposit the particles of carbon, the coarser at a distance from their place of exit and the finest nearer to it.

In order that the process may be carried on uninterruptedly on a large scale, the lampblack chamber should be in duplicate and either should be capable of being put into communication with the chimney separately. It is then possible to be making lampblack in one half of the plant while it is being collected from the other.

Lampblack chambers have been so constructed that the combustion gases with the carbon particles floating in them are compelled to pass through a vaulted brick tunnel, and through an opening at the end of that into a second tunnel running parallel with the first so that the gases travel in the second in the direction opposite to that which they took in the first. This may be repeated with a third and a fourth tunnel before the gases reach the chimney. If each tunnel is eleven yards long and there are four of them the gases have to go forty-four yards. The last tunnel receives the finest lampblack.

In constructing this apparatus it is essential to line the tunnels with good smooth bricks and to point the courses well, so that the lampblack can easily be brushed off the sides. Before the plant is used it must be heated for a considerable time to dry it completely. This is done by passing through hot air and gases from a furnace until the outside of the tunnels is distinctly warm. If this precaution is neglected the lampblack produced will be worthless for a long time. The heat of the gases will evaporate the moisture out of the walls and it will again condense in part upon the particles of lampblack so that instead of a dry loose powder the product will be a smeary black clinging mass which will have to be dried and ground before it can find a market. It is easy to see that the plant described costs money and can only be recommended for working on an unusually large scale, and a plant which will yield two cwt. and more of lampblack every twenty-four hours can be erected in a much cheaper way.

For the purpose of burning the raw material for lampblack, a brick building is erected about eleven yards wide, eleven to thirteen yards long, and ten feet high. The interior is plastered smoothly inside with ordinary mortar and divided by a brick wall longitudinally into two chambers, each eleven to thirteen yards long and about sixteen feet wide. At the end opposite the hearth is the chimney which can be connected with each of the two chambers in turn. The height of the chimney must be at least eleven yards, and on the top of it there must be a damper to be presently described for regulating the draught.

Figs. 31 and 32 show the arrangement: Fig. 32 in ground plan, and Fig. 31 a vertical section. Both the chambers are divided into compartments by walls parallel to the width of the chambers and about forty inches apart. It will be clearly seen from the figures that the combustion gases entering at F and F_1 must work their way round these inner walls, so that they travel sixty to seventy yards in each chamber before they escape into the chimney. The inner walls subdividing each chamber should be of iron plate, preferably galvanised except those near the chimney, which may be of wood planed smooth. It is sufficient to have the first two or three in each chamber of metal to prevent any risk of ignition by the heat of the combustion gases. Each chamber is provided with two doors one close behind the other. The outer door is an ordinary wooden one, which should, however, be lined with sheet iron as a protection against fire. The inner door is of wood and lined with thick felt where it meets the doorpost to make the door shut as air-tight as possible. No draught should be able to pass into the chamber except through the same opening as the combustion gases, round the inner walls and out at the chimney *s*. To enable the lampblack to be removed from the chambers by daylight, the side walls have windows one between each part of the inner walls. These windows are forty inches long and eight inches wide, and are most

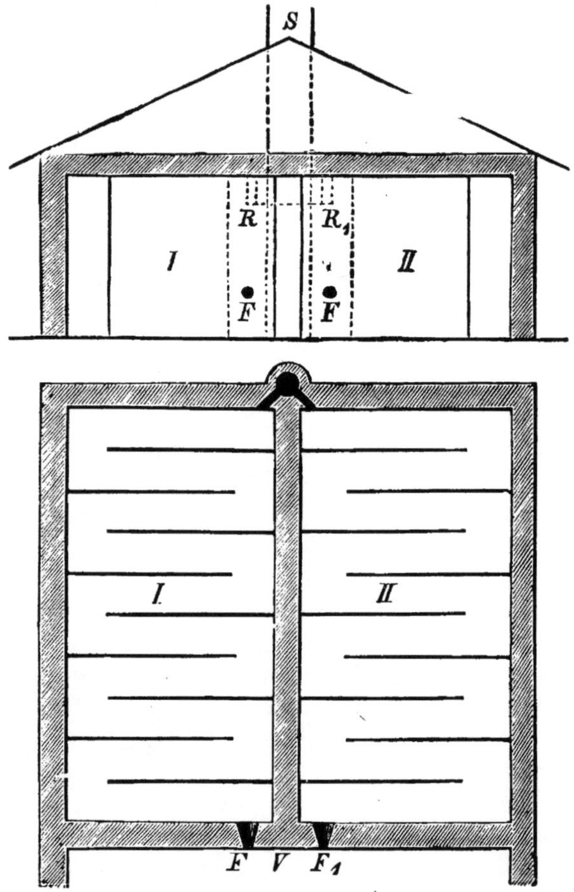

Figs. 31 and 32.

easily made by fixing sheets of glass air-tight by means of cement into openings in the brickwork. The windows may be dispensed with and the lampblack collected by artificial light. In this case no candle or lamp should be used, for fear of sparks igniting the lampblack, but an electric light carried by the workman together with a small accumulator.

Where the gases pass from the last subdivision of each chamber into the chimney are dampers R and R_1 whereby each chamber can, if required, be shut off entirely from the chimney. To regulate the draught with sufficient accuracy a damper acting with perfect certainty must be put at the top of the chimney. Such a damper is shown in Fig. 33.

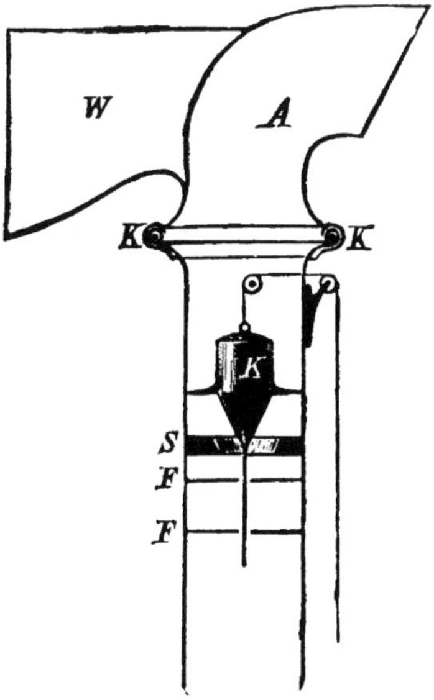

FIG. 33.

The iron cowl A rests on the chimney and bent so as to open horizontally, and is balanced by the iron plate W. The cowl runs on a ball-bearing K, so that it acts as a weathercock and adjusts itself so that the wind can never blow down the chimney and hinder the proper movement of the fire-gases. This arrangement works perfectly even in the windiest weather.

Below the cowl is the damper for regulating the speed of the gases in the chimney. An iron plate S with a circular opening in the middle of it fits into the chimney. The opening is wider above than below, so that it can be closed by the heavy iron cone K. K is guided by an iron rod which passes through central holes in the cross rods F F, and is raised or lowered by a chain passing over the two pulleys seen in the figure. Fig. 34 shows the lower end of the chimney shaft.

The channels I and II lead one from each of the two chambers K K into the vertical part of the shaft. At R and R there are accurately fitting dampers above each of which is a Bunsen burner (G G_1). If one of these dampers is opened and the burner above it lighted, the air in the channel is heated and rises up the chimney,

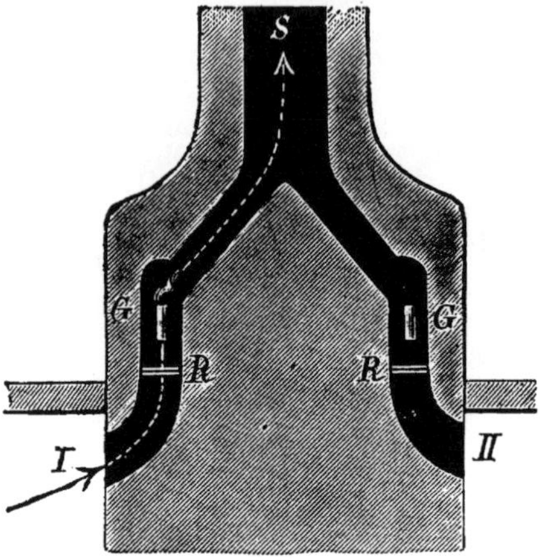

FIG. 34.

thus causing a draught in the chamber and causes the combustion gases to pursue the frequently described devious course in the chamber.

THE BEGINNING OF THE PROCESS.

When a newly erected plant of this description is used for the first time, there are usually various difficulties to contend with which gradually disappear as time goes on. These difficulties are principally two in number, one is that instead of getting a pure light dry product a blackish brown greasy black is formed which lumps together when taken from the chambers and can be made no use of unless first thoroughly dried. The other difficulty is that it is hard at first to hit upon the exact regulation of the draught necessary at the same time to bring all the combustion gases into the chimney and not to impair the smoky character of the flame. If the draught is too strong it carries away fine particles of carbon, and instead of the just perceptible smoke which ought to be seen issuing from the chimney there is a thick black cloud containing a large proportion of the best and finest of the lampblack. If on the other hand the draught is too weak, all the lamps may go out on account of the accumulation of the products of combustion in the chambers as they cannot escape by the chimney. Hereby the atmosphere in the chambers becomes incapable of supporting combustion and the lamps perforce go out.

It often takes a long time to get the apparatus to work properly, without anyone knowing why they did not do so from the first. In many cases the reason is attributed to defective construction and still oftener to want of skill on the part of the workman. Neither supposition can be justified. If the building were defective the arrangement could never be got to work properly, but would have to be rebuilt, and an attendant who would never learn the simple physical and chemical processes concerned in the manufacture of lampblack would never succeed in making a faultless product. He would in every sense of the word drive money out of the chimney, as it is much easier to work with a strong than with a weak draught, although as above stated much black is swept out at the chimney so that the output is far less than when the work is properly executed.

Before setting a new plant to work it is necessary to consider whether it is suitable for the purpose. Every manufacturer will agree that when the building is complete and fitted with all the accessories for work, it should be started at once, for all delay means a time during which the capital expended on it is lying idle.

Obvious as this may be it does not quite hit the mark with a new lamp black chamber. However sparing the bricklayers may have been with their water in mixing their mortar, the new brickwork must be very damp and the walls will feel damp and cold. Left to itself the brickwork dried but slowly, and even in a hot year it takes months before the work is apparently dry. Even then, if it is heated it will at once be evident that it is not really dry but only on the outside. The wet from the inside will begin to sweat out, and this goes on until the brickwork is dry all through. It is therefore essential to wait till the brickwork is quite dry before setting the apparatus to work. If this is not done the product will, for a long time, be damp and clinging instead of dry and loose. The only way to save the months of waiting while the brickwork is drying spontaneously is to dry it artificially. The expense of doing this cannot be compared with the loss of interest on capital involved in the long waiting, for they are confined to the cost of a few cwt. of coke.

The drying must not be done without care. If the brickwork is heated too fast it may crack. The heating must therefore be gradual, but must be kept up uninterruptedly till everything is quite dry. Great care must be taken, too, that no ashes from the fire get into the chambers or the first lots of lamp-black will be, if not altogether useless, at least of very inferior quality. The drying is best done as follows:—

Build in the lamp-room when the black-producing material is to be afterwards burnt a cylindrical brick oven about five feet high and twenty to twenty-four inches in diameter. In this, place a grate about four inches from the bottom and connect the oven near the top with an ordinary stove pipe which must be provided with a damper. The top of the oven is closed with a removable iron lid. The stove pipe is led with a slight upward slope to one of the openings through which the lamp-black is to pass into the chambers, and is luted into it air-tight. The oven is then filled with coke, and while this is being done the damper must be quite shut to keep coke dust out of the lamp-black chamber. The damper is then opened, as well as that governing the bottom of the chimney, and the damper at the top of the chimney is opened full. Straw is then lit in the chimney to make a draught through the lamp-black chamber. When this has been set up small pieces of soft wood are lit under the grate in the oven so as to heat the lower part of the coke red hot. The heat soon spreads till all the coke is glowing and turning gradually to carbonic acid. This gas enters the lamp-black chamber at a very high temperature, and loads itself with moisture from the brickwork until steam begins to appear at the top of the chimney. When the coke in the oven is nearly burnt out, the stove-pipe damper is shut, the oven is quickly recharged and the damper is again opened. The heating must be kept up continuously day and night until the brickwork is quite dry, as shown by the clouds of steam ceasing to issue from the chimney, and the outside of the brickwork feeling warm even in those parts farthest from the stove.

The lamps can then be put in their openings at once and the action started as there is already a draught through the chambers and the chimney.

It may happen even with a perfectly dry lamp-black chamber. which has been at work some time, that the lamp-black comes out damp and clinging instead of dry and loose.

The reason must now be in defective working as the brickwork has long been dry. In the imperfect combustion taking place in the lamps with a limited supply of air, the products contain not only carbon separated out in the solid form and also certain volatile tarry bodies, but always in addition certain quantities of the normal products of organic combustion, carbonic acid and steam. Now the draught must always be strong enough for this steam to retain the gaseous form till it escapes from the chimney. If the draught is too weak some of the steam will condense on the floor of the lamp-black chamber, and in sinking on it will carry down particles of lamp-black and form with them a damp cohering mass. Too much draught is wrong because it sweeps lamp-black out of the chimney. The loss of lamp-black is certainly least when the draught is too weak, but entails the formation of a damp product, which is a very great inconvenience. It is hence of the utmost importance in lamp-black making to hit upon the happy medium as regards draught, and the ability to do this can only be acquired by practical experience. Practical men take for their guide in this matter the appearance of the flames. These should have a cloudy red colour, and their tips should be bent over with streaks of lamp-black going from them to the openings into the lamp-black chamber. The appearance of the flames in conjunction with the look of the products escaping from the chimney top really constitute a good test of how things are going.

Occasionally a sudden break in the satisfactory working occurs. The lamps previously burning in quite a normal manner, cease to do so, and the draught has at once to be altered. This is always caused by a very quick change in the barometric pressure, as can be proved by consulting a barometer at intervals. If the barometer suddenly falls, the draught becomes too quick, and too slow it the glass rises suddenly. Such sudden changes in pressure often occur in a few minutes as precursors to thunderstorms or gales. When therefore either is impending extra attention should be given to the lamp-black chambers. In starting a new apparatus for the first time, begin with one chamber. Open the damper at the top of the chimney completely, and also that governing the bottom of the chimney, and light the Bunsen. The damper of the other chamber must be closely shut. Bring from time to time the flame of a candle to the opening in the back wall of the chamber where the combustion gases enter it. The flame should bend to the opening as this is a sign that a draught has been established through the whole apparatus. The manufacture of the lampblack can then be started. The combustion products are allowed to enter the chamber and a constant watch is kept upon the top of the chimney. The gases pass rather fast round the partitions in the chamber so that only a small part of the lamp-black suspended in them can settle upon the floor. Hence as soon as they begin to issue from the chimney the fact is announced by the formation of a thick black cloud of smoke.

As soon as this makes its appearance the Bunsen is put out so as to lessen the draught, and the smoke from the chimney becomes thinner. After this it is a question of so regulating the draught that the velocity with which the gases pass through the chamber to the top of the chimney becomes small enough to allow the particles of lampblack to settle to the floor. To ensure this the damper at the bottom of the chimney is gradually closed, as well as that at the top, until the smoke at the chimney top is only just perceptible. When this is the case, the speed of the draught is as small as it can be with safety. As soon as one chamber has been properly started in this way, the second can be set to work in the same manner. It is, however, better to wait till the first chamber has been going for twenty-four hours before starting the other, so that when it is fairly under weigh the lamp-black can be removed from the first. To stop the first chamber extinguish the soot-producing flame, and let the chamber cool for a few hours. During this time the draught through it gets weaker and weaker, and all the lamp-black settles, while the chamber gets full of air coming through the opening which formerly admitted the combustion gases. The damper at the chimney foot is then entirely closed and the lamp-black is taken out. It can be sorted while doing so as the finest particles are those found nearest the chimney. The lamp-black is best scooped out with enamelled iron buckets with their bottoms covered with india-rubber or felt to preserve the cement floor of the chamber from injury. The man who removes the lampblack must take care not to take any dirt or grit into the chamber with him on his boots. To keep the lamp-black clean he should always wear felt shoes kept for this purpose only. The chamber is usually emptied in thirds, one consisting of the part nearest the chimney, one that furthest from it, and one between, so as to get three degrees of fineness. Lamp-black will be found on the partitions as well as on the floor. This is usually left alone, as it gradually falls on to the floor in subsequent usings of the chamber. If however, a chamber is to be disused for some time, this black is swept down with a soft broom and removed. As even with the greatest care it is impossible to prevent some of the fine lamp-black rising in dust, the collector should never enter the chamber without having his mouth and nose covered by a respirator so save his lungs from the effects that any inhalation of the dust would produce. A mask such as firemen wear at fires answers still better. This mask consists of a helmet enclosing the whole head, and provided with a valve opening outwards to permit respiration. At the back the helmet is connected with an india-rubber pipe of such a length that the man is free to go to any part of the chamber outside which the pipe is connected with a small condensing pump to supply him with fresh air. He therefore only breathes the pure air from outside and cannot be injured by the lamp-black dust.

The longer the chambers can be worked day and night the better, and the work of each chamber should not be stopped for any longer time than is absolutely necessary to get the lamp-black out of it. Such uninterrupted working gradually heats all the brickwork in a very uniform manner, and this makes it far easier to regulate the draught properly.

The working of the chambers must be constantly watched. If in burning oil, for example, a few more lamps are started, the increased development of heat will strengthen the draught as will be shown by too much smoke coming from the chimney. The draught must consequently be weakened by using the dampers. When a new plant has been worked for a time, sufficient practical experience of its idiosyncrasies will have been gathered to enable the attendant to adjust the draught quickly until the sign that all is right is observed—viz., a very thin grey smoke at the top of the chimney.

Whatever sort of lamp-black is wanted the lamp-black chambers are always built the same. The only difference is in the construction of the combustion room. Resins are burnt in shallow pans, rosin oils on red hot plates, and fatty oils and petroleum in specially constructed lamps. We now proceed to describe these different combustion arrangements in detail.

Resin Lamp-black.

It would be possible with a proper combustion chamber to make lampblack by the imperfect combustion of ordinary pine-wood, but the product is always of small value, so that the method is never used in a properly conducted factory. But a fine lamp-black can be got from purified resin, and large quantities are made in that way.

The hearths for burning resin in a limited supply are usually very simple in construction.

One of them is shown in Fig. 35. A niche O is made in the wall of the lamp-black chamber and is closed by the iron lid D, so as to form a space the shape of half an egg. The floor of this space consists of the iron dish G, surrounded by a water-jacket to keep G from getting too hot. The flame from the burning resin goes through the opening R into the chamber. It is easy to see that this arrangement is very imperfect. When the lid D is raised to put more resin into G, that left begins to burn with a bright flame, and the draught in the lamp-black chamber is considerably increased, and, besides, it is impossible to regulate the draught properly by means of D.

FIG. 35.

Fig. 36 shows a somewhat better contrivance for controlling the draught.

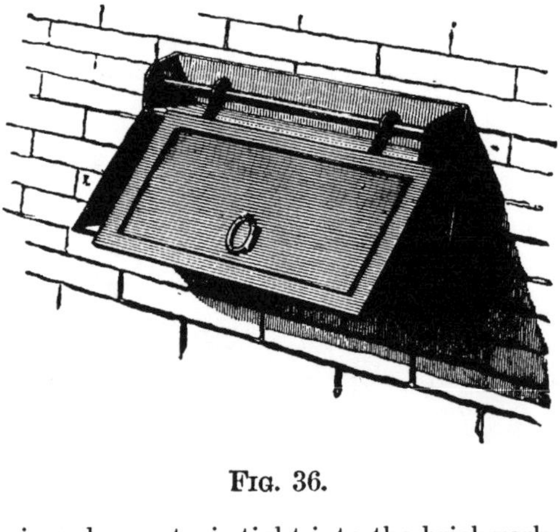

FIG. 36.

This is an iron box set air-tight into the brickwork. The front of this box is cut aslant and the oblique opening is provided with an iron lid sliding end-ways on a rod, so that a slit can be opened for the admission of air into the combustion room. As this opening can be made of any desired width a means is afforded of exactly regulating the draught. In burning resin for lamp-black care must be taken to allow as much heat as is requisite to burn all of it. It is true we get more resin with less air, but as resin oil is formed it must be burnt or it will contaminate the lamp-black and convert it into a greasy sticky mass. The flame will show whether enough air is being admitted to burn the resin oil. This is the case when the flame is a dull red and reaches completely to the opening into the lamp-black chamber. In order to be able to observe the flame without opening the damper too far, the movable iron lid of the box is provided with a circular hole glazed with a piece of mica by way of a window. Through this the flame can be watched without altering the slit admitting the air for combustion. The lid need then be lifted only when all the resin is nearly burnt and a fresh charge is necessary. As this, however, disturbs the adjustment of the draught, it is preferable so to arrange matters that the fresh charge of resin can be introduced without moving the plate. A section of a suitable arrangement is shown in Fig. 37.

FIG. 37.

Partly embedded in the wall separating the combustion room from the lamp-black chamber is an iron pan H filled with rosin. The heat from the rosin burning in the dish S keeps the rosin in H melted. When it is seen through the mica window A in the sideways movable cover D that the rosin in S is nearly exhausted, the valve V in the bottom of H is lifted by the lever shown in the figure until the necessary quantity of rosin has run into the dish. With this simple contrivance as much rosin as required may be run into S without ever altering the opening of D, once the draught has been properly adjusted, and the process can be kept going for days on end without anything being done but to keep H supplied and to run the rosin into S from it from time to time. The contrivance has also the advantage that it can be used not only for rosin but for all other employable solids, so that pitch and petroleum residues can be used just as well. Whatever material is used, the chief thing is to manage the combustion in such a way that no liquid distillates contaminate the lamp-black and make it greasy.

Making Lamp-black from Liquids.

The chief liquids used for the manufacture of lamp-black are rosin oil, liquid animal and vegetable fats, crude petroleum, the camphene or protogene got in rosin-distilling, and now from the competition of other products hardly able to find a market as an illuminant, and finally a number of volatile hydrocarbons obtained in refining petroleum and known collectively in commerce as crude petroleum ether. Petroleum-ether, however, is now much less used than formerly for lamp-black manufacture, as on account of its use as a motor spirit its price has much increased, and is now almost too great for our purpose, although the ether yields a very valuable lamp-black. The price of vegetable oils, too, shows a constant tendency to rise, and it looks as if they, too, would soon have to be erased from the list of lamp-black materials. Among oils, fish and seal oils appear to be the only ones which we shall be able to use much longer for lamp-black manufacture. They are still to be had fairly cheap, especially as only the cheapest unrefined kinds are needed for the purpose.

On considering the methods of using liquid fuel for lamp-black many great differences will appear between them, as oils differ very much among themselves. Fish oil, petroleum, and vegetable oils, can be burnt in lamps so that their product is really "lamp" black, but though petroleum—ether and other volatile hydrocarbons can be burnt in lamps, they are usually treated by a different method.

Rosin oil, an exceedingly important material for lamp-black manufacture, has too high an ignition point for it possibly to be burnt in lamps, and is also too thick to rise properly in wicks. The same remarks are true of some petroleum residues. Properly to utilise these bodies for lamp-black making it is necessary to heat them to make them thin and to let them fall then upon red hot plates. On touching a metallic surface which shows at least a dull red heat, they partly evaporate and partly decompose into less complex hydrocarbons which then evaporate also. If the proper quantity of air is then allowed to come into contact with the combustible mixture of gases, they burn with a cloudy reddish yellow flame, from which a large quantity of extremely finely divided lamp-black deposits.

Making Lamp-black from Rosin Oil.

In making lamp-black from rosin oil or thick petroleum oils, the apparatus represented in Fig. 38 renders excellent service.

Fig. 38.

It consists of a closed iron vessel with a vaulted top, and connected by a pipe with the lamp-black chamber. The front of this vessel is closed air-tight while it is at work by a lid in the upper part of which is the mica window *a* for observing the flame. Below the window the lid has a circular hole inside which is a hoop bearing a screw with a rather fine thread. An iron disc *b* is screwed on to the thread outside. As the disc has a rather greater diameter than the hole it can be screwed up so as completely to close the opening, or may leave a narrow inlet. With it, therefore, the supply of air to the burning oil can be accurately regulated as the opening can be adjusted with great delicacy, as the thread of the screw is fine. The bottom of the iron vessel is corrugated, and about 8 inches above it is a circular tube. In this there is a hole above each of the furrows in the corrugated bottom. The ring tube is connected by a wider tube with the bottom of an iron pan in which the oil to be used undergoes a preliminary heating. The conical valve at the bottom of the pan is adjusted by a screw, and allows the flow of oil to be exactly regulated.

When the apparatus is to be used, the pan is filled with rosin oil, and a fire is lit under the corrugated bottom of the domed vessel. When this bottom is red hot, the conical valve is raised a little, so that the rosin oil falls in separate drops into the grooves in the red hot bottom. As soon as it touches this, the rosin oil begins to decompose and evaporate. The disc regulating the draught is then adjusted so that each drop burns as it falls with a cloudy flame, the point of which sends out a black streak of smoke towards the pipe leading to the lamp-black chamber. It is easiest to regulate the combustion so as to get the most lamp-black, when each drop of the rosin oil can be seen separately as it falls. Each drop must be fully burnt before another falls from the same hole, and the conical valve must be adjusted accordingly. By working in this way not only is the combustion properly regulated, but the formation of coke in the grooves is almost entirely prevented. If the oil drips too fast, there is no time to evaporate it completely, and some of it forms a coke-like mass in the grooves. This coke not only represents a waste of carbon, but by being a bad conductor of heat, makes it necessary to burn far more fuel than would otherwise be required to keep the grooves red hot. Nevertheless, it is impossible altogether to prevent the formation of this coke so that the apparatus must be allowed to cool at intervals in order that the grooves may be cleared out with a suitably shaped tool. For this purpose the front plate is removed and the pipe to the lamp-black chamber is completely shut with an iron tompion. This must be done to prevent the strong draught which would otherwise ensue in the chamber. When the grooves have been cleaned, the tompion is removed, the front plate is replaced, and the apparatus can be re-started immediately.

The lamp-black deposited by this system in the lamp-black chamber must be in very fine flakes and be perfectly loose, and not lump together on pressing. If it does lump it is a sign that the grooves were not hot enough to convert all the oil into gas, so that vapours of undecomposed oil went into the chamber with the lamp-black, and condensing there upon it, made it greasy and clinging.

Making Lamp-black from Liquid Fats.

From all the liquid fats which are fluid enough to be burnt in lamps a product can be obtained which is actually lamp-black. It is of special excellence, and is used for the very finest printing ink. It fetches a price which compensates for the expensiveness of the raw material.

It is well known that the lamps in which liquid fats or petroleum are burnt for illuminating purposes are exclusively designed to give a perfectly smokeless flame, giving as much light as possible. With this object the wick is made cylindrical so that air passes inside it as well as outside, and a tall glass chimney is used to create a strong draught. The result is that the carbon separates out in a state of very fine division from the oil, and at a white heat gives a strong light. It is finally completely burnt so that nothing issues from the chimney but carbonic acid and steam. The better the lamp is for lighting, therefore, the worse it is for lamp-black making. A lamp for this purpose must be fed with a strictly limited supply of air, so that only a feebly luminous flame arises from the wick and gives a black cloud of unburnt carbon. A lamp which fulfils these conditions very perfectly is that of J. Bersch, shown in Fig. 39.

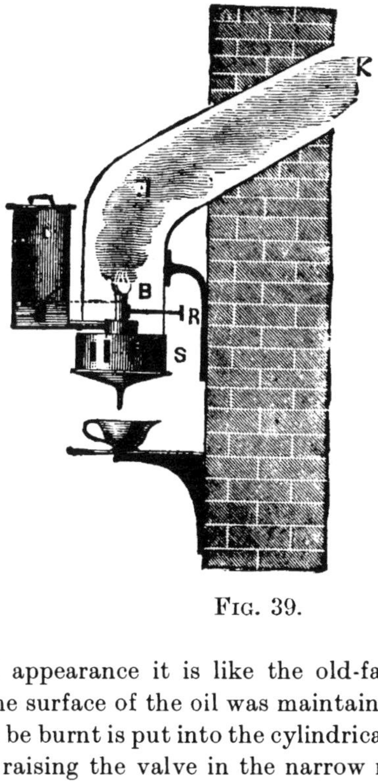

FIG. 39.

In its outward appearance it is like the old-fashioned reading lamps in which the surface of the oil was maintained at a constant level. The oil to be burnt is put into the cylindrical vessel C, which is first closed by raising the valve in the narrow neck. C is then placed in the vessel N. If the short wire to which the valve is attached touches the bottom of N, the valve is open and the contents of C flow out. But as soon as the oil rises sufficiently in N to close the neck of C, the flow ceases. The wick of the lamp must be turned so as to be level with the surface of the oil in N. When some of the oil has been burnt the surface of it in N sinks sufficiently to let some more oil out of C, until the neck of that vessel is again below the surface. Hence the wick is fed with oil until C is empty. The wick is wide and flat, and enclosed in an iron tube H, bent as shown in the figure, and opening into the lamp-black chamber K. Small mica windows should be inserted opposite each other in H, whereby the appearance of the flame can be observed. The screw R regulates the height of the wick.

On the lower part of the tube H a short cylindrical piece S is fitted. This ends below in a cone, the apex of which is perforated with a small hole. Below the hole is placed a vessel to catch the oil which flows over from the lamp unburnt. As the figure shows, the circumference of S bears a number of rectangular slits which correspond to exactly similar slits in H. When the lamp has been lighted the draught is adjusted by turning S, so that the slits in it overlap those in H more or less, until the flame of the lamp is a cloudy red with a copious cloud of smoke. The height of the wick is regulated so that the flame is as large as possible consistent with its having these properties, and the draught is regulated accordingly. At first, however, the flame should be very small, and should not be increased until the tube H has been well heated. The little flame carries all the combustion gases in the still cold apparatus into the lamp-blackchamber, but if the flame is large from the first, large flakes of soot will collect in the bend of H, and finally fall upon the lamp. Once, however, H is hot enough, the speed of the gases in it is sufficient to prevent any deposition of carbon until the chamber is reached.

As every lamp must be put out of action from time to time, to clean it or to put in a new wick, the lamps must be capable of removal, together with the cylindrical piece S. While the lamp is away, the lower end of H is kept plugged by a stopper kept handy for the purpose. This is to prevent a strong draught going through it into the lamp-black chamber.

As many lamps are attached to a lamp-black chamber as there is room for. The attendant who has to superintend the combustion and keep the lamps supplied with oil, has so much to do that he may be unable to watch all the flames properly, and so cause a loss of lamp-black. It is therefore extremely advisable to adopt

Bersch's suggestion to supply all the lamps from a single oil reservoir. When this is full, the man has nothing to do for a long time but to devote his attention to the separate flames, and regulate the draughts. Fig. 40 shows the automatic contrivance for feeding the lamps.

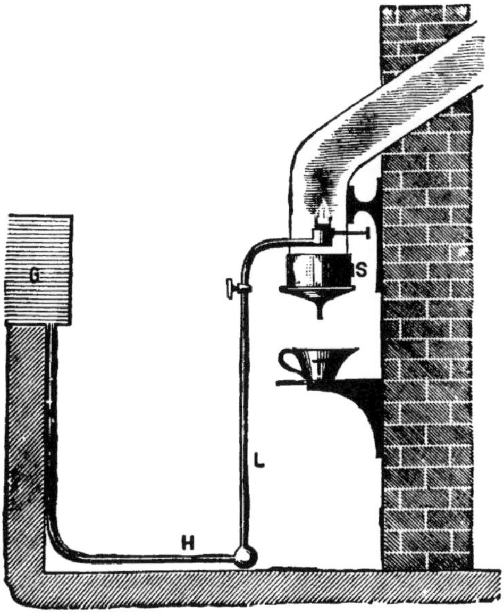

FIG. 40.

The large oil reservoir G is made large enough to keep all the lamps going for at least twenty-four hours. It is placed higher than the lamp-wicks. From its bottom proceeds the tube H which communicates with another running along the floor of the lamp-room. From this proceed other tubes L, one to each lamp. A cock in each of the tubes permits the regulation of the flow to each wick. As the common reservoir is very large, a long time elapses before the surface of the oil in it has sunk so low that any diminution of the hydrostatic pressure becomes noticeable. The cocks in the tubes L must then be opened a little wider to admit the necessary amount of oil to the wicks. It is better, however, to arrange the common reservoir, exactly as we have described is done with the separate reservoirs, and to supply it from yet another reservoir. In that case the level of the oil in it cannot change as long as there is oil in the second reservoir.

UNIVERSAL FEED FOR THE LAMPS.

To provide any given number of lamps uninterruptedly with fuel, the best arrangement ever proposed is the one which can be called by the above name. It is so constructed that the oil reaches the separate wicks at a perfect uniform rate, provided (and this is essential) that the tops of all the wicks are at exactly the same level.

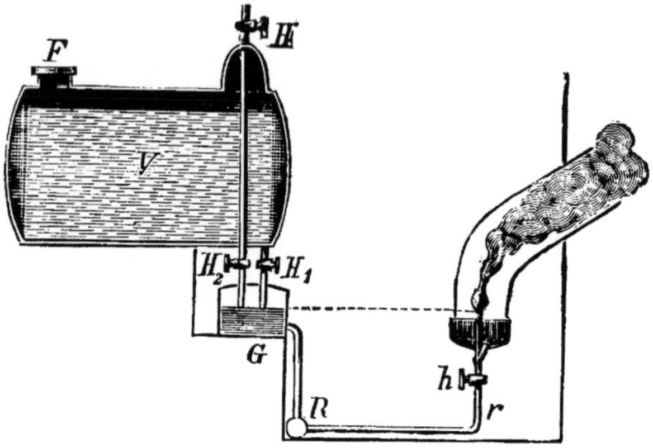

FIG. 41.

Fig. 41 shows the apparatus. The oil reservoir V is of sheet iron and completely closed. It may be of any size desired. It is filled at F, through an opening provided with a lid, which can be screwed on air-tight. There is also an air-tight cock H. From the bottom of V issues a short tube governed by the cock H_1, and leading into the closed receptacle G. Parallel with this tube is a narrower one H_2, which is continued into V and opens freely above the surface of its contents. A pipe goes from the bottom of G to a wider horizontal tube R, from which proceed the small tubes *r* to the separate lamps. V is filled through F, the cock H being open, till the oil has reached the bottom of the small dome into which H opens. F and H are then shut and H_1 and H_2 are opened. The oil then flows through H_1 into G, the air previously in this escaping through H_2 into V. As soon as the oil is deep enough in G to close the lower ends of H_1 and H_2 it ceases to flow from V, and cannot start doing so until it has sunk so far as to leave both tubes open again. Then the small cocks *h* are opened to admit the oil to the lamps, and when these have been lighted the supply to them is adjusted by giving the cocks *h* the proper degree of opening. If at the same time the draught is set so that they burn with a very smoky flame, the attendant need not trouble about the lamps any longer while the oil in V lasts, so that one man can manage a large number of lamps, as all he has to do is to keep the wicks at the right height. As the arrangement just described excludes the possibility of the fuel burning, except in the lamps, it can be used for such liquids as petroleum, which catch fire without any wick, on contact with a flame.

If the apparatus is to be used for burning petroleum, it is superfluous to use lamps at all, and a simple arrangement is substituted in which the petroleum burns freely. This arrangement consists of a horizontal pipe, which is fixed to the supply pipe inside the lamp, and has on its upper surface a row of holes large enough to admit a good-sized pin. The taps *h* are turned so that the petroleum slowly oozes from these holes. It is then set fire to, and should burn in separate flames each about four-fifths of an inch in height. This is easily managed by means of the taps *h*. The lamp case is then closed and the draught is arranged so as to make the flames as smoky as possible.

It has also been proposed to use an automatic contrivance, whereby the petroleum to be burnt flows at the proper rate into small shallow dishes. These are enclosed in a casing whereby the draught can be regulated and the combustion gases led into the lamp-black chamber. It is obvious that this is a possible method, but comparative experiments have shown that the yield of the lamp-black is greater when the petroleum is burnt drop by drop as above described, and also that the lamp-black is got in a much drier state.

This result is due to the fact, that when the petroleum is burnt in dishes, some of it is always volatilised unburnt and carried into the amp-black chamber. This serious drawback may be partly avoided by keeping the dishes cool during the combustion by means of a water-jacket, but the method of burning by drops is always to be preferred.

Making Flame Lamp-black.

Every luminous flame, whether of a candle, or of a lamp fed with oil or petroleum, or of coal gas, consists of two very distinct parts. The lower part consists of gases comparatively poor in carbon, which burn with a pale, usually light, blue flame. The upper part, which runs to a point, consists of the glowing products of the combustion of these gases, *i.e.*, of carbonic acid and steam, in which very minute particles of solid carbon are floating, having separated out from the flame gases, rich in carbon. These particles of carbon are carried upward by the glowing gases, and if there is air enough, are gradually burnt, so that nothing rises from the apex of the flame but carbonic acid and steam. In this case complete combustion takes place.

If the luminous part of the flame is cooled in any way, as by holding in it a cold piece of metal or porcelain, the particles of glowing carbon are suddenly chilled. They can then no longer burn, and are deposited on the cold surface in the form of a very delicate deep black powder, *i.e.*, in the form of lamp-black.

Lamp-black so obtained has many advantages over that made in other ways. It appears perfectly dry, has a pure black and consists of nearly chemically pure carbon. It owes this high degree of purity to the manner of its formation, which involves the absence of nearly all the tarry substances that are formed in the other processes, and cling to the free carbon.

Such carbon, deposited by cooling a flame, may be called flame lamp-black, and many forms of apparatus have been constructed for preparing it. Independently of the superior quality of the flame lamp-black, the use of these apparatuses offers the great advantage that large and costly erections are not needed even to produce large quantities of lamp-black. Indeed a number of such apparatuses can be set to work in an ordinary living room, and will make considerable quantities of lamp-black in a single day. Fig. 42 illustrates the principle of all these contrivances.

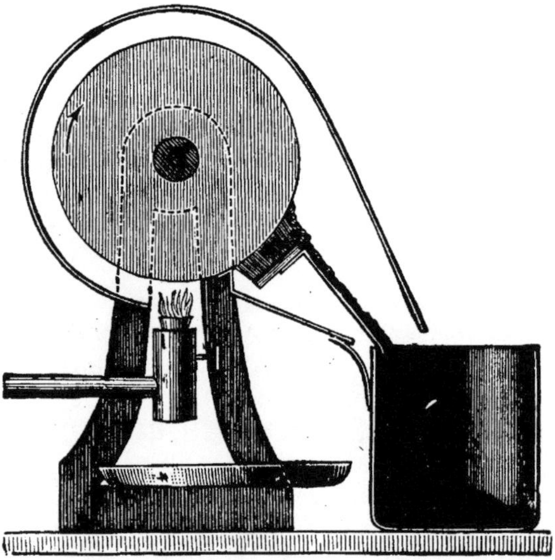

Fig. 42.

A cast-iron cylinder with thin walls is turned smooth outside. It can easily be rotated on horizontal hollow trunnions, through one of which water is admitted to the inside of the cylinder, and flows out out again by a pipe from the other. This constant flow of water will keep the outside of the cylinder quite cold. Round the cylinder, and an inch or two from it, is a sheet of iron casing, the form of which will best be gathered from the figure. Under the cylinder is a row of lamps, and their flames strike upon the cylinder. This, being kept cold by the flow of water, becomes coated with flame lamp-black. A very soft brush is fixed, so that as the cylinder rotates it brushes the lamp-black from it and guides it into a receptacle. The cylinder is turned by any suitable mechanism, but the motion should be slow, so that the layer of lamp-black is not formed too thin. This apparatus will make lamp-black from animal or vegetable oil or from petroleum. When petroleum is used it is advisable to put under the rotating cylinder two or three parallel horizontal tubes having numerous small holes at the top, from which the fuel issues in small drops. The tubes should be arranged in such a way that the holes do not form rows running at right angles to the length of the cylinder, or the lamp-black will not be distributed uniformly over the cylinder.

An extremely fine lamp-black can also be made by burning ordinary coal gas from the holes in the tubes, but as a rule coal gas is too dear in towns to be an economical source of lamp-black. The case is different with rosin gas, which can be very well utilised in this way, and it gives a lamp-black fine enough for making the very best printer's ink.

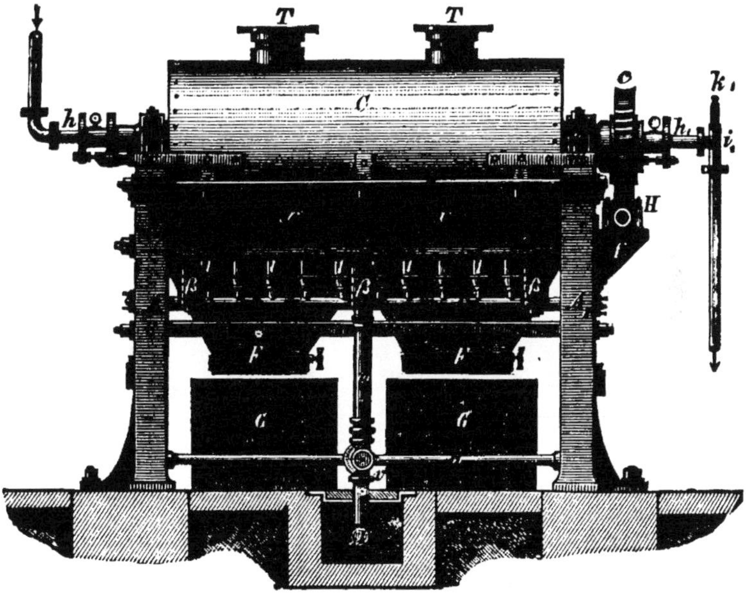

FIG. 43.

An apparatus constructed on the principle shown in Fig. 42, and one which fulfils every proper requirement for the production of large amounts of lamp-black, is that of Dreyher, shown in Figs. 43

to45, inclusive. Fig. 44 shows the lower part of the apparatus from the outside, Fig 45 a longitudinal section, Fig. 46 a side view, Fig. 47 a cross section, Fig. 48 a section of the man-hole and of the counter weight to the cylinder B, and Fig. 49 an upper view of the footplate of a side-wall at the framing AA. Fig. 50 gives a side view of the

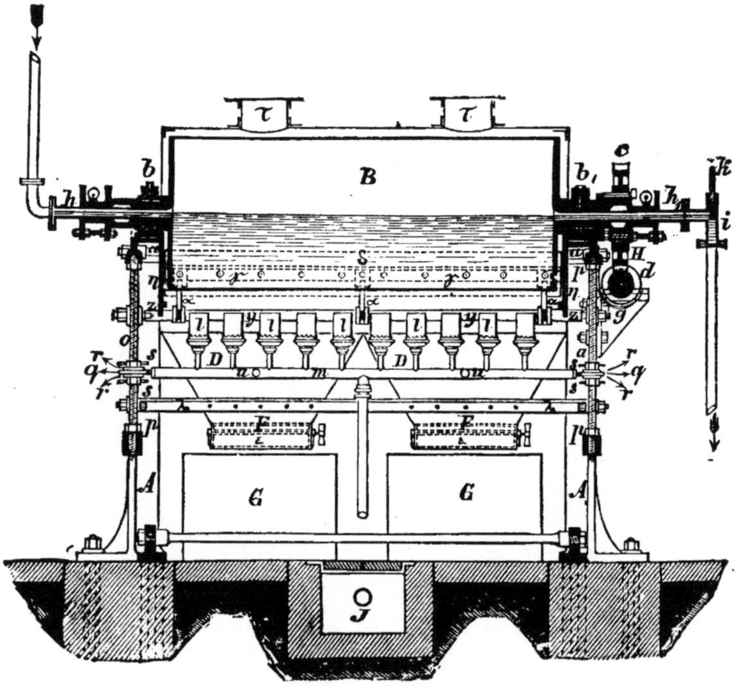

FIG. 44.

same, Fig. 51 a longitudinal section, Fig 52 a cross section, Fig. 53 a view from above. In Fig. 54 a section through the lamp-black collector *a* is given, and Fig. 55 shows the arrangement of the cam *n* and the lifting rod *o*.

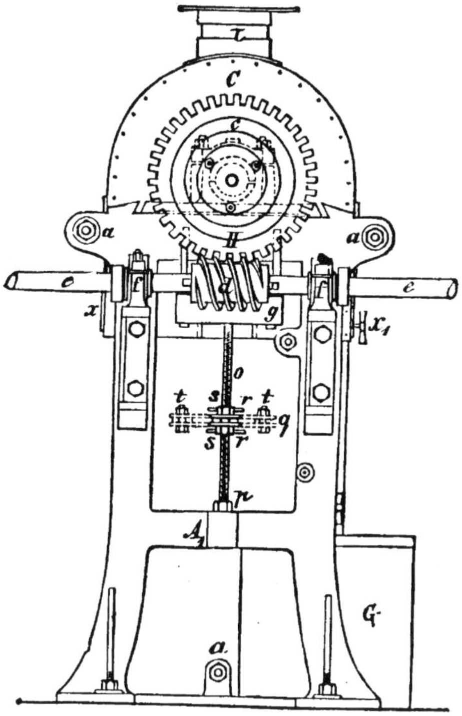

FIG. 45.

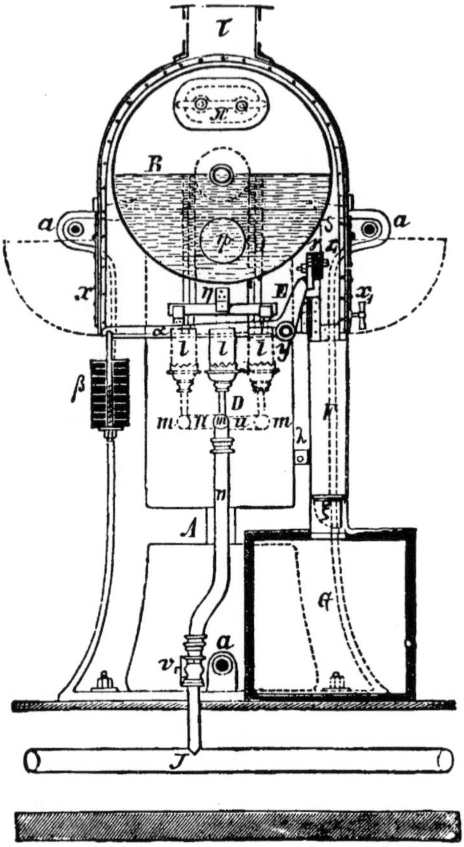

FIG. 46.

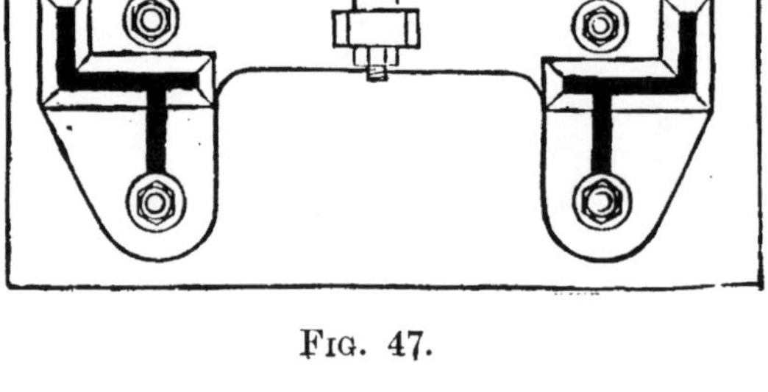

FIG. 47.

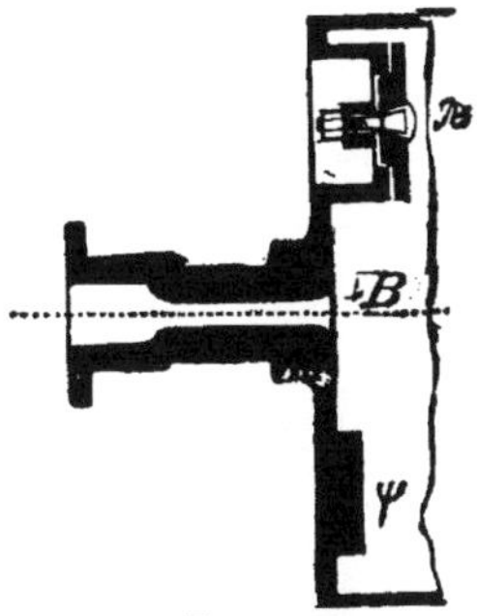

FIG. 48.

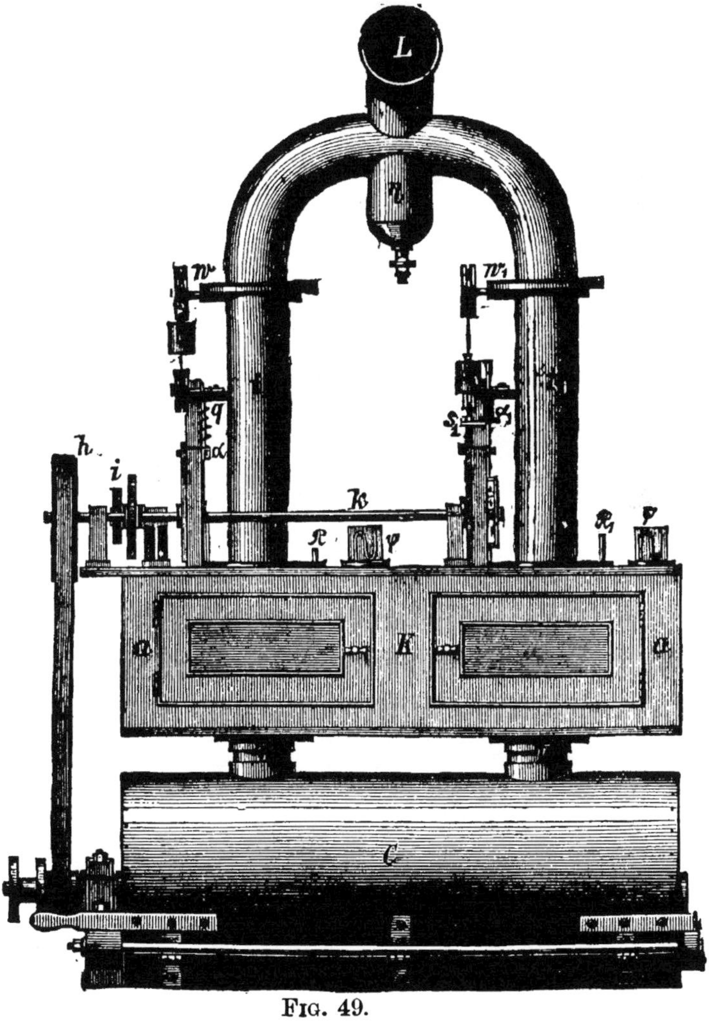

FIG. 49.

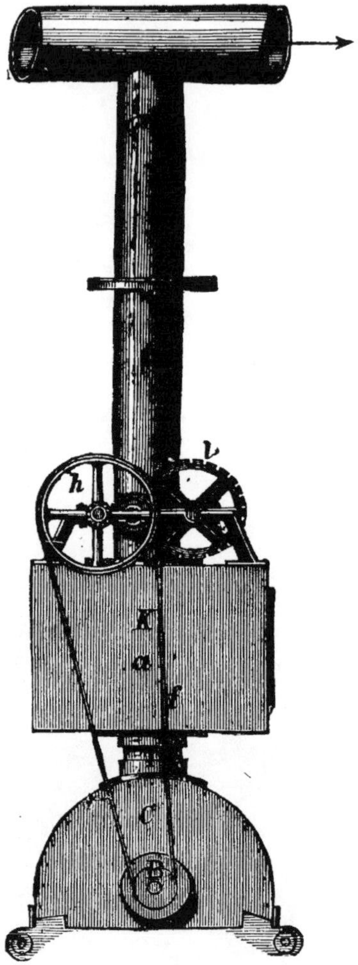

FIG. 50.

FIG. 51.

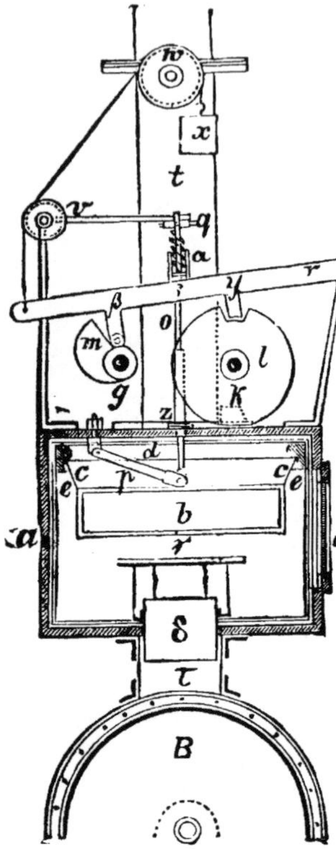

FIG. 52.

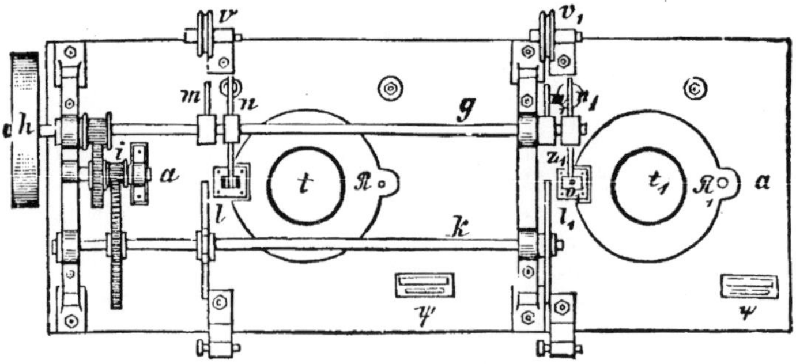

FIG. 53.

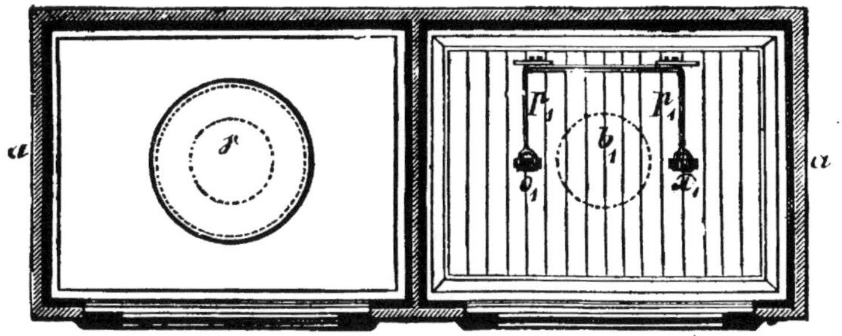

FIG. 54.

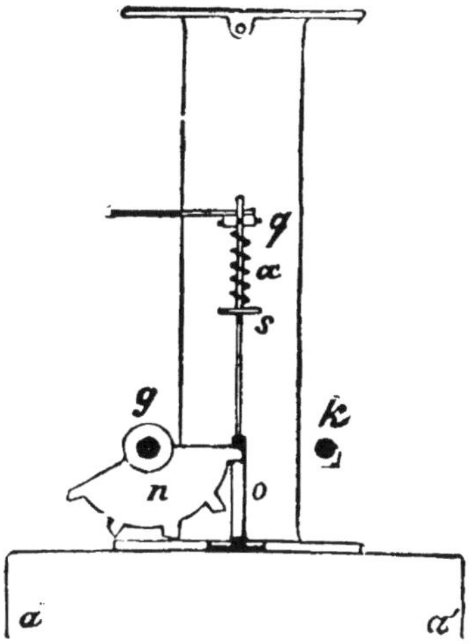

FIG. 55.

The separate parts of the apparatus are as follow:—The frame AA, the cylinder B, the casing C, the lamp-guard D, the stripper E, the collecting funnel F, the gearing H, and the aspirator K. The frame is composed of two strong cast-iron ends, connected by the screws *a*. In the bearings *b* and b_1 rests the polished cylinder B with hollow trunnions, to one of which the worm wheel *e* is fixed. This is geared with the endless screw *d* and rotates the cylinder. Water flows from a reservoir through the cylinder in the direction of the arrow and leaves the apparatus warm. To determine exactly the flow of water necessary to keep the cylinder cold enough a thermometer *k* is inserted at *i*. When the lamps are burning the flow of water is regulated so as to keep *k* always at the same temperature.

The number of lamps used depend on the size of the whole apparatus. The lamps *l*, and the distribution tube *m*, and the rubber tube, are connected with the mechanism intended to bring the lamps to such a distance from the outside cylinder, that the flame is so cooled as to give the greatest possible amount of lamp-black. Two round iron rods *o* are stuck into holes in the frame, and fixed there by the nuts *p*. At the closed ends of the distribution tubes, forks *g* are placed, and lie between the plates *r*, which are fastened to the round rods by the nuts *s*. As the figures show, three rows of lamps can be used with the apparatus and burnt altogether or one or two rows at a time.

The tube J supplies fuel to the lamps, and it may be any vegetable or fish oil or petroleum, etc. The supply is regulated by the cock *v*, which is connected with the distribution tubes, and the lamps by the india-rubber tube *n*.

The casing C of sheet iron is an inch or two away from the cylinder B. It can easily be lifted off by the handle *w*. At its lower end are two flaps, *x x*, one giving access to the lamps, while the other allows the action of the stripper to be observed.

The stripper is essentially a strip of thin elastic steel δ which is fixed by screws on to the rails η η. By loosening the screws the stripper can be pushed along the rails, so as to adjust its distance from the surface of the cylinder B as may be desired. The aspirators K are connected with the upper part of the casing, and are connected with the outlet pipes *t t*, which are themselves connected with a ventilator. The draught of air caused by the ventilator is intended to free the combustion gases, which consist chiefly of carbonic acid and steam, from the lamp-black still floating in them and to lead them to the chimney.

The most important part of the whole aspirator arrangement is the lamp-black catcher. This is a wooden box divided into two compartments, and covered outside with felt to prevent it from cooling too fast. Each compartment contains a lamp-black collector consisting of a sheet of iron cut zig-zag, and covered with coarse flannel. On the rough surface of the flannel the lamp-black settles as it is carried upwards by the combustion gases. This would soon choke the lamp-black catcher, were it not for a device by which the lamp-black is constantly being shaken off the flannel. As will be seen from the figure, the flannel-coated irons are lifted at intervals and allowed to fall again, whereby the lamp-black is knocked off on to the bottom of the box. To prevent the lamp-black thus shaken oft from falling again upon the cylinder B, guard lids *y y* and insets δ δ are provided, under which the combustion gases pass freely through the supports to the wooden box.

Dreyer has also fitted his apparatus with various other accessories for regulating the rate at which the cylinder B rotates, etc.

As will appear from this description, the Dreyer machine is very complicated, and, like all such machines, requires unremitting attention while in action, or very soon much trouble will be incurred. Nevertheless, as large an amount of lamp-black as desired can be made with very simple apparatuses, two of which will now be described.

Fig. 56 shows Thalwitzer's apparatus.

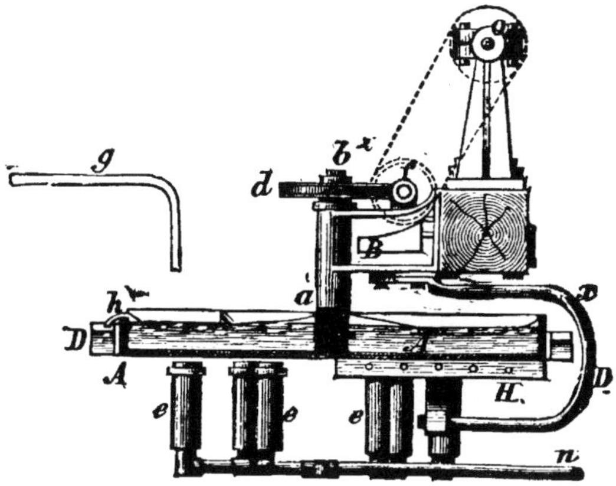

FIG. 56.

The chief part of this is an iron plate A, having a bent-up edge *a* and resting upon a vertical axis, which can be rotated. The gearing for effecting this consists of the cast-iron *b*, the cogwheel *d*, the worm *f*, and the driving shaft *o*. The lamps *e* are put under A, and are connected by *n* with a fuel reservoir common to all of them. H is a stripper consisting of a slanting plate of steel, which continually scrapes off the soot settling on the under side of A when the apparatus is at work.

When the lamps are lighted and the distance of the flames from the underside of A has been properly adjusted, the rotation is started, and water is run on to the plate A from the tube *g*. As soon as the water has filled A up to the top of the bent-up edge, it overflows into a circular gutter which surrounds the plate.

Fig. 57 represents an arrangement very similar to the last.

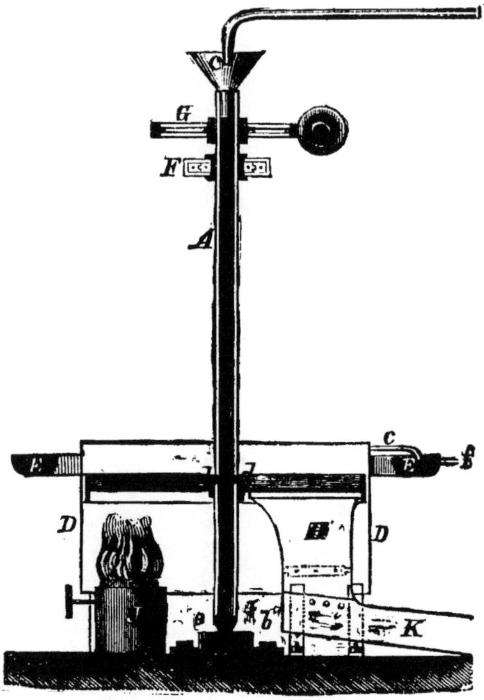

FIG. 57.

The water for cooling flows in this apparatus through the openings *d d* in the hollow axle on to the plate, and after cooling it flows off at *e* by the circular gutter *f*. The lamps J and the stripper H are enclosed in an iron casing D.

PURIFICATION OF FLAME LAMP-BLACK.

Flame lamp-black, by whatever method it has been prepared, has certainly a deep black appearance to a cursory glance, but a practised eye will at once observe that the black is not pure, but that the colour is a deep, blackish brown. This colour appears most distinctly when a mark is made with the lamp-black on white paper, close to another made with some purified lamp-black. The crude lamp-black will then show distinctly brown, while the other streak will be pure black in the thicker parts, fading to grey at the edges.

As it is impossible so to conduct the combustion of the bodies used for lamp-black manufacture, that nothing is formed but pure carbon, steam, and carbonic acid, the lamp-black is always contaminated with products of dry distillation. These are solid or liquid at ordinary temperatures, and sometimes these bodies are present in rather considerable quantities, and are the substances which impart the brown colour to the lamp-black. Hence, before pure black printing inks can be made from it, the lamp-black must undergo a very careful purification. This can be done by two methods, by calcination at a red heat, or by dissolving out the impurities from the lamp-black by reagents. The simpler of the two processes is calcination in closed vessels. The high temperature either volatilises or cokes the tarry products adhering to the carbon. The process, however, differs in no respect from that formerly described for the purification of ordinary pine lamp-black, and requires only a few words. In carrying it out it is of the greatest importance to raise the temperature very gradually, watching the flame issuing through a small opening in the lid of the calcining vessel. If this flame is very long and bright, and comes out with a whistling noise, the heat is too great, and there is some risk that the pent-up gases will blow the lid off, whereupon the lamp-black inside would at once catch fire and be destroyed. The flame must be small and silent, and the fire must be kept at that point till the flame disappears entirely. The fire is then increased until the vessel is at a bright red heat, so as to conclude the process quickly.

To determine the action of very high temperatures on lamp-black we have subjected it for a long time to a bright white heat. The result was a velvety black substance, which proved to be nearly pure carbon on analysis. When this lamp-black was worked up with oil in a paint mill, it was found that it required much longer rubbing up with the oil than lamp-black which had been purified at a red heat only. This remarkable fact can only be explained by the heat coagulating the lamp-black, and, considered in connection with the greater consumption of fuel for getting the higher temperature, shows that a red heat is best for practical purposes.

A properly calcined flame lamp-black answers every requirement that can be made in respect of pure black colour, and softness of the colours, so well that it can be used for the finest inks for art printing. Experience has however shown that engravers on stone and copper give the preference to chemically purified lamp-black, and this cannot be a matter of prejudice, because whether the engraver knows how the lamp-black is made or not he makes this choice.

This is proof positive that the chemically purified lamp-black has advantages as regards softness and colour which escape any but the experienced eye.

Chemical Purification of Lamp-black.

All the tarry matters which produce the brownish shade are fairly easily soluble in alkalis, both caustic and carbonated, especially the former. Hence it is advisable to purify lamp-black from these substances by means of fairly strong caustic soda lye used hot, and to repeat the washing several times.

It will be observed in testing the process on a small scale that the first boiling with caustic soda lye gives a very dark brown solution, the second a slightly paler liquid, but even the fourth liquor is fairly dark. This result is due to the very great colouring power of the tar-products, so that even traces of them give a very distinct colour to the lye. Quantitative experiments have shown that the great majority of the tar-bodies are dissolved out by the first boil, and that there is very little even in the second liquor. Hence two boils are quite enough in practice to get as good a lamp-black as can be wanted for any purpose.

The boiling of the lamp-black with caustic soda lye is done in iron pots, and the following process can be recommended as using the least possible quantity of alkali.

The lamp-black is put in the pot, and the caustic soda lye, containing 15 per cent. of Na HO (34° Tw.), is poured in so as to soak it and just cover it. Water is then stirred in until the mass has the consistency of thin cream. It s then boiled up. As soon as it is boiling hard cease heating and let most of the lamp-black settle. The dark brown supernatant liquid is then ladled out into a second pot on to more crude lamp-black. When this has been boiled the lye is regarded as exhausted. In the meantime a fresh lot of lye is put into the first pot, but this time a lye of 5 per cent. (12° Tw.) only. After boiling, the lye is drawn off and the lamp-black is rinsed with water and dried It is then fit for any purpose that may be required of it, even the most exacting use to which lamp-black can be put.

The weak lye after it has been used for the second boiling of the first lot of lamp-black is used for the second boiling of the next, and is used on in this way as long as there is any lamp-black to purify.

The lamp-black retains quantities of lye, and must be very thoroughly rinsed to get rid of it. The rinsing is done in shallow vessels having a bottom full of holes. Over these holes is spread a closely woven linen cloth over which the lamp-black, wet with lye, is uniformly spread. Here it is left until lye ceases to drop through and the surface of the lamp-black begins to crack. The lamp-black is then covered with water and again allowed to drain perfectly. This process is repeated three or four times, and then nothing remains but to dry the lamp-black. Some insist that the rinsed lamp-black should be boiled with nitric acid and again rinsed, but this further treatment appears unnecessary. Nitric acid is expensive, it must be used in stoneware or porcelain vessels, and its action does not in fact add to the purity of the lamp-black. Treatment with caustic soda lye alone will give a product which on paper makes a perfectly black mark without any trace of brown.

Care must be taken that the rinsing is complete before drying, or else white efflorescences of carbonate of soda will appear in the dry lamp-black. To see whether the rinsing is complete take a little of the lamp-black off the linen, filter and dry it at a gentle heat in a porcelain dish. The lumps of carbon left must fall to dust at a touch. The dish is then heated over a flame till the lamp-black catches fire and gradually burns away. If the rinsing has been sufficient, nothing but the merest trace of ash will be left on the dish. In the contrary case, there will be a distinct white residue of carbonate of soda.

The lamp-black is dried by spreading it thinly on linen cloths stretched on a wooden frame. The frames are then put into a drying room, heated from the outside. The temperature inside the rooms should be about 40° C. This temperature will dry the lamp-black very quickly if the room is well ventilated and kept clear of the steam rising from the lamp-black.

It is a very bad plan to try to hurry the drying by doing it for example on hot metal plates. There is great danger when this method is adopted that the carbon, being in a state of minute subdivision, will catch fire and be destroyed. If this happens, the only plan to save anything is to throw all the lamp-black at once into a vat of cold water. When the lamp-black is dry, it is left on the frames till the room is quite cool, and it is then at once filled into the receptacles in which it is to be sent out, which should be iron canisters with well fitting lids. When the canister is full and the lid on, a strip of paper is gummed round the edge of the lid to prevent either air or dust from getting at the contents.

In working at the purification of lamp-black on a large scale, considerable quantities of used-up caustic soda lye rapidly accumulate. This lye is dark brown and has a peculiar smell, but the caustic soda can be recovered from it with ease by evaporating to dryness in a shallow cast-iron pan. The mass is then treated red hot with constant stirring with an iron rod. The heat destroys the organic matter dissolved out from the lamp-black. The residue is pure carbonate of soda. This is dissolved in ten times its weight of water and the solution is boiled up with slaked lime. When the carbonate of lime formed has settled completely, the clear liquid is drawn off and is a solution of caustic soda, which can be used to boil some more lamp-black. This recovery of the caustic soda can, however, only be done remuneratively by those dealing with very large quantities and having cheap fuel at their disposal.

XIV. MAKING LAMPBLACK PIGMENTS.

The manufacture of lamp-black pigments in general, *i.e.*, printing inks, is carried out in a similar manner to that in which ordinary paints are prepared. The solid pigment, in this case the lamp-black, is mixed in the proper proportions with a drying oil by machinery until a perfectly homogeneous mass is formed. The homogeneity of the mixture is of greater importance with printing ink than with any other pigment. A printing ink which has not been properly mixed produces at those places where there is too much lamp-black and too little oil a deep black wide compression, and at those where there is too much oil and too little lamp-black a grey and narrow mark. Hence it is impossible to use such an ink for printing purposes, especially for art printing. Some copies will come out deep black and with wide lines, and others grey with narrow lines. If the very finest and best purified lamp-black is examined under the microscope, it will be observed that the pigment is not, as might have been expected, in particles of equal size and shape, but in masses showing great differences in both respects. The larger masses consist of aggregations of the smaller ones which cohere together with some force, so that pressure on the cover glass will break up the lumps only imperfectly. Long narrow flakes will also be seen, and are the remains of the large flakes seen under the microscope in unpurified lamp-black. Both the lumps and the flakes will inevitably cause unevenness in the appearance of the finished printing.

Paint Mills.

In order to obtain printing inks conforming to the requirements above indicated, the only means available must, of course, be adopted. It is absolutely necessary to grind them in a paint-mill until all the particles visible in the field of the microscope appear of the same size.

There are many types of paint-mill, and although there is a

number of differences in detail, they can all be classed under three well-marked categories. A cone mill is represented in Fig. 58.

FIG. 58.

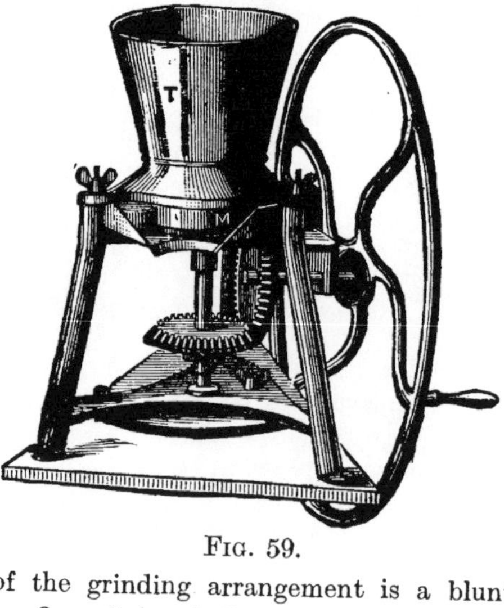

FIG. 59.

One part of the grinding arrangement is a blunt cone with a fluted surface. Over it is a hollow cone into which the other fits.

and the distance between the two can be regulated at will. The grinding is effected between the two conical surfaces. Fig. 59 represents a plate machine.

The grinding is here done between two horizontal plates provided with radiating ribs. In some cases one plate is fixed while the other turns. In other machines both turn, but in opposite directions. The roller machines form a type illustrated in Fig. 60.

FIG. 60.

The rollers are of steel, porcelain, or hard stone. They are truly cylindrical, and polished. A special arrangement permits the exact regulation of the distance between the rollers, which also move laterally as they rotate. Hence the mass passing between the rollers is rubbed as well as pressed.

A machine of any one of these types will serve to produce homogeneous paints, if it is constructed with sufficient care, but for making printing inks the roller machine is the one to be preferred. These turn out the mass in a proper state more quickly than any of the others, but in working on a large scale it is advisable to give the lamp-black and oil a preliminary mixing with a cone or plate mill, and to complete the process in the roller mill.

ROLLER MACHINES.

The best roller machines are those which have a large number of pairs of rollers arranged step-wise one over another in such a way that the mass scraped by the stripper from the uppermost pair passes through the second pair, and from them to the third, and so on The separate pairs are so geared together that the lower rollers move somewhat faster than the upper ones, and that at every rotation one of each pair of rollers moves a little end-ways, and then back again.

The distance between the rollers of each pair must be capable of exact adjustment by means of fine threaded screws. As the lower rollers have to be closer together than the upper ones, there is a tendency for the mass to stick at them. This is prevented by running them faster than the upper rollers, so that the mass passes along uniformly.

The material at present used for the rollers is hard stone or porcelain. Stone rollers are made of porphyry or granite, but the stone must, in any case, be a very fine grained one Experience has shown that both coarse grained porphyry and coarse grained granite are harder in some places than in others. Hence the stone wears faster in the soft than in the hard parts. The latter are, therefore, brought into relief. Such a want of uniformity in the surface, even if imperceptible to the eye, is enough seriously to affect the proper working of the machine. Nothing can then be done but to turn the rollers on the lathe, or to put in new ones,

Porcelain rollers consist of a thoroughly uniform mass at least as hard as porphyry or granite, Being uniform in structure they wear evenly and become even better after long use than when new, as small irregularities of the surface left by the turner get smoothed away, Hence roller machines with porcelain rollers are certainly to be preferred to machines with stone or metal rollers.

TINTED PRINTING INKS.

Printers' inks consisting solely of purified lamp-black and vehicle give, of course, impressions which are pure black. It is, however well-known that a black which has to a practised eye a tinge of blue in it looks much better than a pure black. To make such a tinted ink many makers mix the lamp-black with a blue pigment, which is added in very fine powder before the first grinding. Prussian blue is the pigment usually chosen and gives very attractive results. Prussian blue is, however, not a remarkable stable substance, and is very apt to turn brown from the formation of ferric oxide. Hence an ink made with Prussian blue, although it may look very fine at first, often assumes a dull brown hue in the course of time. Excellent substitutes for Prussian blue are to be found in the Induline blues. These are very fast dyes, and inks tinted with them do not change colour. As pure indigo is now made artificially and sold at a reasonable price, this extremely fast dye can also be used for tinting inks made with purified lamp-black.

Carbon Pigments of other Origin.

Lamp-black is made either by burning bodies containing a large percentage of carbon in a limited supply of air, or by suddenly cooling the flame emitted by such bodies when on fire. In both cases a part of the carbon is separated out in a very finely-divided state as a black powder, containing distillation products from which it must be purified as above described.

There are, however, other ways of getting finely-divided amorphons carbon. One is a process known as dry distillation. This enables a large number of bodies, otherwise of very small value to be used in the preparation of a product which will bear comparison with the finest flame lamp-black in every respect. The cost of carrying out the process, too, is very much less than in the case of flame lamp-black, so that it deserves the special attention of all lamp-black manufacturers.

If any organic substance containing a considerable proportion of carbon is enclosed in a vessel provided with a small orifice for the escape of gas, and if the vessel is then heated red hot, the oxygen and hydrogen of the organic body combine with certain quantities of the carbon present and form liquids of a great variety of composition and inflammable gases. The high temperature at which these liquids are formed vaporises them, and they escape with the gases through the orifice. If the body heated contains a great deal more carbon than the oxygen and hydrogen can combine with it, much carbon remains behind in the solid state, and can be purified for use as a pigment.

If the temperature is gradually raised, the physical structure of the body heated remains quite unchanged. If wood is treated for example, the resulting charcoal will show the microscopic structure of the original wood with perfect distinctness. The organic bodies which can be used for the purpose always contain more or less mineral matter or ash which would remain behind even if the matter were heated with free access of air. The resulting carbon must, therefore, be purified from these bodies by chemical methods whereby the inorganic bodies are made soluble so that they can be washed away with water.

The number of materials available for carbonising is fairly great, and certain waste products for which it would be difficult to find any other use answer very well, The bodies mostly used are scraps of ivory, bone, leather, and, as regards vegetable matter, vinestalks and other comparatively soft vegetable matter. Wine or beer yeast forms a very valuable material, which is as yet but little used, The same may be said of dried blood from the slaughter houses.

Bone or Ivory Black.

All bones, and ivory is bone in a sense, consist of a framework of crystallised matter or bone-earth, in the interstices of which organic matter is embedded. Hence if bones are heated red hot in a closed vessel, the organic matter is destroyed, leaving carbon in a very finely divided state lodged in the bony framework. If the heat is applied gradually the bone retains its shape, but is quite black and of much less weight than at first. This bone black or animal charcoal is a substance which has great power of absorbing colouring matter from liquids, so that it is largely used for bleaching such liquids. For example in the vast industry of beet sugar manufacture, the solutions first made are very dark in colour, but after filtration through animal charcoal will give colourless crystals on evaporation. Chemical trades require such large quantities of bone charcoal that its production is a large industry in itself. As in breaking up the charred bones a fairly considerable amount of waste is produced, in the form of dust and small grains which cannot be used for bleaching purposes this waste should be worked up into a pigment. This is done by dissolving out the mineral in it with hydrochloric acid, and then rinsing and drying the carbon.

The mineral basis of bones consists mainly of the phosphates of lime and magnesia, salts soluble in not too dilute hydrochloric acid. A vat is half filled with the above mentioned waste, which is then just covered with a mixture of equal volumes of commercial hydrochloric acid and water. As the mineral matter also contains carbonates a lively effervescence at once ensues, and small quantities of hydrofluoric acid are also formed from the decomposition of calcium fluoride in the bones. Now hydrofluoric acid is a very dangerous substance, as air containing even traces of it is very injurions to the lungs. Hence the addition of hydrochloric acid should be done in the open air and the vat should be left by itself until the evolution of fumes ceases. A plug is then pulled out of the bottom and the carbon is thoroughly drained. It is then stirred up with water and again drained, when it has fully settled to the bottom. This rinsing with clear water is repeated till all the hydrochloric acid is washed away and only pure carbon remains in the vat. As for pigment making purposes it is essential that the carbon should be as finely divided as possible; it is as well to grind the washed carbon in an ordinary colour mill. Very little power is required for this purpose, as when once the bone-earth is removed the carbon particles have little cohesion. The properly ground mass forms a deep black mud which can be left to dry or be dried by artificial heat. When dry, the purified bone-black is of a pure black and makes a most excellent pigment.

Bone-black is put upon the market under all sorts of names such as ivory black, *ebur ustum*, Frankfort black, neutral black etc., etc. These all consist of finely ground bone-black purified from mineral matter. If leather scraps or dried blood are to be worked up iron tubes are employed, closed at one end, and with a well-fitting lid with a small hole in it at the other. As these bodies give off large volumes of combustible gas during the charring, it is a good plan to lead the vapours from the hole by a bent tube so that they can be burnt and help to supply the heat required and so save fuel. Leather or blood gives a charcoal which hardly requires treatment with hydrochloric acid, for the amount of mineral salts present is so small that its removal appears superfluous.

Vine Black.

The one year old and barely lignified shoots pruned systematically from vines are used in many places for the preparation of a black pigment which from its origin is called vine black. The shoots to be carbonised are packed, cut in equal lengths, into iron pipes which are crammed as full as possible. The plug is then put in the pipe, which is slowly raised to a full red heat. The charred matter contains considerable quantities of carbonate of potash, from the organic potash salts originally present, and is freed from them by washing with water. It is not necessary to treat with hydrochloric acid, as the insoluble salts are very small in quantity. It is true that the product would be of even better quality were the acid used, but water washing and grinding usually constitute the entire preparation.

Soft shoots of other plants can also be carbonised, and give a product which can be used as a pigment after purification. Shoots of porous white wood, such as poplar, alder, and willow, are especially suitable.

Yeast Black.

In the fermentation of grape-must a considerable quantity of a pale brown substance separates out. It consists of the cells of the yeast plant mingled with a large number of small shining crystals of tartar. As this thick yeast still contains about half its volume of wine it is pressed to recover this. The residue must either be worked up at once or thoroughly dried for keeping. Wet wine yeast putrifies with extreme ease, and the putrefaction rapidly destroys its most valuable constituent, the tartar. The dry yeast is carbonised in tubes similar to those used for the calcination of lamp-black. The charred mass is put into a vat of cold water, where it soon falls to a fine powder. When this has happened stir well for a time, and allow to settle, finally drawing off the liquid from above the sediment of carbon.

The tartar which separates out with the yeast during the fermentation consists chiefly of the tartrates of potash and lime. These salts are decomposed by the calcination and converted into the corresponding carbonates. Carbonate of potash is very soluble in water, so that it will be found in the water drawn off from the vat just mentioned. The carbonate of lime is insoluble in water, and hydrochloric acid must be used to get rid of it. Washing with water will then give an unusually finely-divided carbon of very fine colour.

Both dried and freshly pressed yeast are used in special chemical factories for the preparation of tartaric acid. This is done by treating the yeast with hot hydrochloric acid, which dissolves the lime and potash salts, leaving the organic matter of the yeast unaltered. The liquid is then filtered and the yeast remaining in the filters is rejected as worthless. It ought, however, to be dried and carbonised, and would then give, after mere washing with water, a pigment of very superior quality. In the factories where tartaric acid is made from yeast, expense is incurred in getting rid of the yeast freed from salts, and therefore the advisability of using it as above described will be at once manifest.

In breweries and spirit distilleries every fermentation produces enormous quantities of yeast, far more than can be used as such. At present the best use that can be made of it is to boil it and mix it with cattle fodder. It contains large quantities of albuminoids, so that this employment of it is quite legitimate. But the quantities of yeast produced in a large brewery are so great that it is hardly possible to use all of it for fodder. The rest might quite well be carbonised and made into a pigment in the following way.

The pasty yeast is freed as well as possible from water by pressing, then completely dried and carbonised in iron pipes. Although the carbon obtained does contain a certain amount of mineral salts especially phosphate of potash, nearly all of them are removed by simply washing with water. After this washing nothing remains to be done but to dry the carbon.

PRINTING INKS.

With the exception of inks required for colour printing, all printing inks are black, and owe their colour to finely divided carbon made from pine-wood, rosin oil, etc., according to the quality of the ink desired. The finest inks of art printing, lithography, and copper-plate are made from flame lamp-black. When we consider the great variety of uses to which printing inks are put and how cheap many of them have to be, the reason for the vast number of published recipes will be very evident. Up to the middle of the 19th century there was but one printing ink, but nowadays there are a great many, and their composition varies considerably. A very great influence on the development of the printing ink making trade was exercised by the spread of periodical literature during the second half of last century. This has caused a gigantic increase in the amount of printer's ink used, and the demand has grown steadily for the last fifty years.

The manufacture of a printing ink depends upon the use to be made of it. There are, however, certain requirements made of all printing inks alike, and these are as follows:—The ink must be a thick and perfectly homogeneous liquid, it must contain no solid matter but finely divided carbon, and every drop when examined microscopically must appear as a clear liquid containing black grains uniformly distributed through it.

The consistency of a printing ink must be such that it passes on to the printing rollers at the proper rate. It will hence be obvious that various consistencies are demanded according to the nature of the machine used by the printer. For a rotatory machine which prints many thousands of copies an hour a consistency will be necessary, very different from that required for art printing or for ordinary work. As regards colour, ordinary printing ink should be a pure black. For economy's sake, however, newspaper printers often use an ink so diluted that it does not look deep black, but a greyish black, especially in large type.

The question of the time that the ink takes to dry on the paper is a very important one, especially with ink used for printing newspapers which are folded and despatched as soon as printed. If then the ink does not dry very quickly, the whole impression smudges and "sets off" so much that it becomes illegible in places.

Although it is essential to have a quickly drying ink for this purpose, it is dangerous to go too far, for a too quickly drying ink would make the paper stick to the formes and tear it. A last condition which must be fulfilled by a good printing ink is that it must be easy of removal from the type, which has to be used again.

It is easy to see from a review of all these considerations that no one composition will answer every purpose and that a number of different inks are required. Makers of printing inks are obliged therefore to work from definite recipes so as to be able to turn out exactly the same ink again and again. They make newspaper ink for rotary presses, book printing inks and art inks. As the recipes have only been attained by long, laborious, and costly experiments, it is obvious that the makers are not disposed to communicate them and the recipes that are offered and published must be looked upon with caution as many of them are of little or no value. In the recipes given below for printing inks, the only intention is to give hints of the general composition and the practical man will easily discover, what, if any, alterations have to be made in the recipe for his special purpose.

XV. MANUFACTURE OF PRINTING INKS.

Many different materials for this manufacture are given in recipes, so many in fact that it is impossible to discover what use they are in the ink. The following is a list of the articles now really in use for the manufacture of printing ink :—

Boiled linseed oil, boiled without driers.
Rosin oil from the dry distillation of rosin.
Rosin itself, especially American pine resin.
Soap, usually rosin-soap, but occasionally ordinary soap.
Lamp-black and various other pigments.

The art of making printing ink remained at a very low stage of development for a long time, and the inks used during the first half of the nineteenth century probably differed little in composition from those which were prepared by the inventors of printing. Linseed oil was very slowly heated over an open fire until it ignited. It was allowed to burn for a time and then extinguished by putting a lid on the pot. In this way a liquid was obtained coloured of a dark brown or black colour with particles of carbon; and with a consistency varying with the period of heating, the thicker, the longer the heating was continued. If necessary, the liquid was then thinned with unboiled, or only very slightly boiled, linseed oil. Lamp-black in the proper quantity was added and the mixture was finally rubbed up on a stone in small quantities at a time to make it uniform. That it was impossible, however, in this way to get uniformity of composition throughout a large mass of ink is best observed by examining old books when it will be seen that whole pages are of a uniform deep black while others are of a greyish tint.

The first real improvements in making printer's ink consisted in the introduction of rational methods of boiling the oil, and of mechanical mixing appliances. The latter were at first ordinary cone or plate colour mills. The roller machine is a more recent date A great stride was made about 1850 when rosin oil began to be used in place of part of the linseed oil and when rosin soap was added to the ink. It then became possible to make inks much cheaper than could be made with linseed oil alone. This was a very important matter for newspaper printing for the amount of ink used in printing, even a single day's issue of a large journal is so great that a saving of only 1 or 2 per cent. on the cost of the ink amounts to a notable sum of money.

Boiling the Linseed Oil.

This process although it goes by the name of boiling is not so in the proper sense of the word, but a heating having for its object an initial oxidation of the oil, so that it will dry better. Linseed oil is a type of the drying oils, those which when exposed in thin coats to the air absorb large quantities of oxygen and are thereby converted into tough solid sheets having properties very similar to those of soft india-rubber. The process goes on much faster with the aid of heat than at the ordinary temperature, and the rate at which the boiled oil will dry in the ink can be exactly regulated by heating it for a longer or shorter time. Prolonged heating gives an oil which will dry very quickly on exposure in thin coats to the air, the shorter the heating the more slowly will the ink afterwards made with the oil dry.

Linseed oil must always be boiled in vessels where it has plenty of room, as the oil soon swells up and it begins to decompose so energetically at a particular temperature that there is considerable risk of its boiling over and catching fire. Various contrivances have been thought out for boiling large quantities of the oil with safety, such as pans with an outlet pipe in the side, through which the oil escapes when it rises too high instead of over the edge of the pan, and fires built on a trolley running on rails so that they can at once be moved from under the pan if there is any probability of the latter boiling over. The best apparatus for preparing thickened linseed oil is undoubtedly one in which the oil offers a very large surface to the air, and on that account requires to be moderately heated only. The oil soon becomes very thick under these conditions and if necessary can be diluted to any required consistency with unboiled oil.

In boiling linseed oil down to the proper thickness by the old method there are two points demanding special attention. One is the liability of the oil to boil over, and the other consists in the development of large quantities of vapour, mostly of acroleine, which have a most powerful and disagreeable smell, and an intense action upon the eyes. The attendant must be protected from these fumes, and the boiling must therefore be done where there is a strong draught to take the fumes as fast as they are produced. There are various contrivances to cope with boiling over. That invented by Andres is shown in Fig. 61.

FIG. 61.

The pan C is suspended by the chain K, by which it can be lifted out from the hearth. A rather large dish-like arrangement R surrounds C and the lid D closes C securely. If the oil boils over it is caught in R. When the oil catches fire, the flame is extinguished by the lid D and the pan is lifted from the fire as quickly as possible, and not replaced till the oil has cooled considerably.

Fig 62 shows an arrangement for boiling linseed oil which not only prevents any injury to the workmen from the fumes of the oil but prevents any fear of loss by the oil boiling over. The pan which can be filled nearly to the brim with the oil to be boiled, has above it an enlargement A which has the shape of a truncated cone and bears the conical head H which is connected to a chimney in which there is a good draught by a wide tube. A rather wide side tube R proceeds from the side of A and is bent over an empty pan K_1.

When oil is heated in this apparatus all the fumes pass through H into the chimney. The shape of A gives full scope for the oil to expand and if even that is not enough, the oil boils over through R into K_1 and is not lost. When the oil has left off frothing, any oil in K_1 can be put back into K and the apparatus simple as it is prevents all risk of fire or loss of oil.

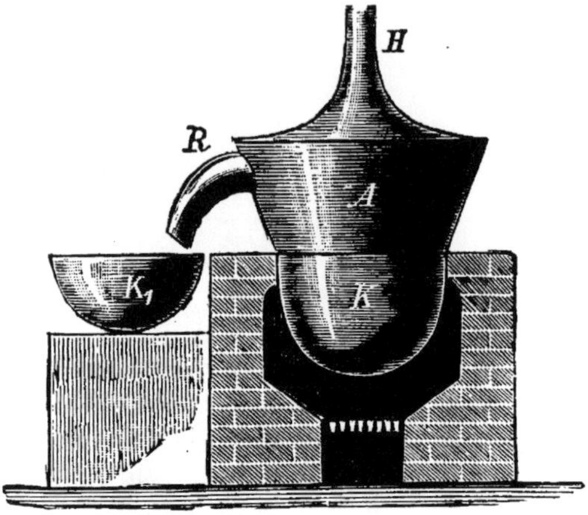

FIG. 62.

The time during which the oil has to be heated depends upon its quantity, on the power of the fire, and on the consistency to which the oil has to be brought. No hard and fast rules therefore can be laid down. The longer the heating continues the darker and thicker the oil becomes and the quicker it dries in thin layers.

TESTING THE BOILING OIL.

To find out the condition of the hot oil at any given moment it is usual to take samples from time to time and pour them in a thin stream into a very thin metal dish floating in cold water. The oil thus rapidly cools to the ordinary temperature and can then be tested. A very favourite test with practical men is to touch the oil in the metal dish with the finger, slowly raise it and the longer the thread which can be drawn from the surface of the oil without breaking, the thicker the oil has been boiled.

A skilled workmen can judge the oil very well in this way, but great experience is required to do so. A far more reliable and just as simple a method is to take the specific gravity of the oil.

The more it has been boiled, the higher its specific gravity. As, however, the indications of an ordinary immersion hydrometer would be unreliable in such a thick liquid as boiled linseed oil, another method which gives very accurate results can be recommended. Take out about a pint of the hot oil with a ladle and cool it quickly by pouring it into a metal dish floating on cold water. Then fill with the oil a metal dish of known weight and holding exactly 500 c.c. when filled to the brim, and weigh the 500 c.c. of oil on a balance indicating a milligramme with that load. The weight in grammes of a litre (1,000 c.c.) of the oil will then be twice the difference between the weight of the full and the empty dish, and the specific gravity of the oil will be one thousandth part of the weight of the litre of oil.

When a number of determinations of specific gravity have been made of the finished oil, the average of the numbers obtained will give a normal specific gravity which can be worked to in all future cases, boiling the oil to a sample with the specific gravity taken as the normal amount. The heating darkens the oil but that is of no consequence with oil to be used for making printing ink, but is on the other hand of very serious importance in making inks which are not to be black, especially for delicate shades in colour printing. In this case only oil of the very palest colour can be used, and it is best not to boil the oil but to thicken it by another process which we now proceed to describe.

Thick Oil.

As above stated the thickening of linseed oil results from the absorption by it of oxygen from the air, which absorption is greatly promoted by heating. There is, however, another way of accelerating the absorption. The oil is only gently heated and is then allowed to fall in a shower through a space in which an upward current of warm air is travelling. An apparatus for carrying this out is shown in Fig 63.

L is a shallow cylindrical dish of metal which can be filled with oil through the opening at O. L is completely enclosed below and at the sides in a second vessel M, but with a space between forming a jacket into which steam can be admitted to heat the oil in L. The condense-water escapes from the opening at E. A pipe coil lies on the bottom of L and terminates in the middle of L in a vertical pipe, the opening of which lies under a conical iron hat H. The end V of the coil outside the apparatus is connected with a fan which drives a continual stream of air through it. The air is warmed by passing through the coil immersed in the warm oil and escapes from under H. Above the vessel L is a vessel G several yards high and made with glass sides so that the drops of oil falling down are exposed to the action of the daylight. Thus the apparatus gives a combination of all the conditions conducive to the oxidation of oil. Great surface, heat, and plenty of air and light.

Over the tall glass case is a reservoir B of about the same capacity as L. This vessel has a perforated bottom such that the oil drops through it as a very fine rain into L. The warm air rising through

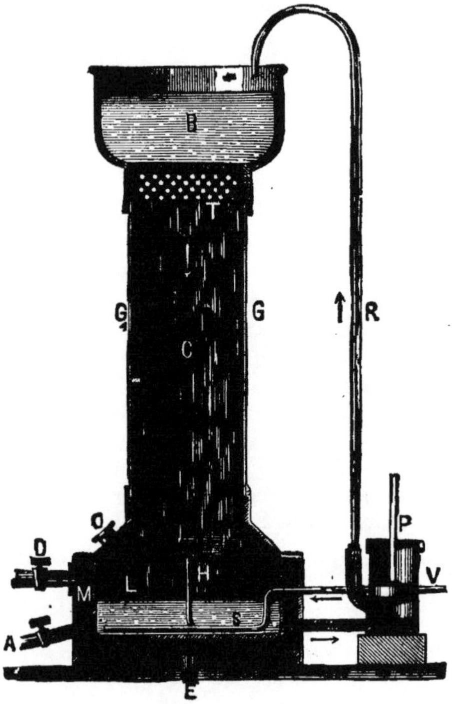

FIG 63.

G escapes at openings just below the upper reservoir and shown at T. The pump P driven by a small motor continually pumps the air back again into B, and the oil is returned to B again and again until it has acquired the desired consistency. In a few hours the oil is generally so far thickened that it spins threads through the holes in the bottom of B, but the time naturally varies with the temperature and the height of the fall. Samples for judging the progress of the process are taken from the filling opening at O. Oil prepared in this way is made only very slightly darker than at first and can be used for the preparation of coloured inks.

Boiled Oil and Rosin.

To make an ink that will not spread at the edges of the letters with nothing but pure linseed oil and pigment it is necessary to use an oil which has been very greatly thickened by one of the processes already described, and the resulting ink, although perfect in its way is very expensive. In making cheaper inks thickeners are added to the drying linseed oil and in this way inks suitable for ordinary printing are made.

The thickeners used are not only rosin itself but rosin soap, and when the ink has not to be of specially fine quality a linseed oil which has been only slightly thickened by oxidation can be used. The process of adding the thickeners is not difficult but require some care as they must be quite free from solid particles.

The rosin is melted in a pan with constant stirring, and then the rosin soap is gradually stirred into it a little at a time, waiting till each lot is dissolved before adding more. The soap must be quite dry and should be added in the form of thin shavings. While the soap and rosin are being mixed, the linseed oil is being heated up in another pan over which is hung a closely woven linen filter. Into this the rosin and soap are poured and carefully stirred into hot oil as they drip through. From time to time samples of the oil are taken and their consistency when cold is noted. If it is too thick it can be diluted with more oil. If it is too thin, the addition of the rosin and soap is continued. It is possible to get a single mixing which will answer every purpose. It is usual to prepare three stock mixtures, weak, medium, and strong. The following recipes show how these mixtures may be prepared, but the proportions may be altered so as to produce intermediate sorts if desired. It will be observed that rosin oil is used in some cases. Like the drying vegetable oils it thickens and dries by absorbing oxygen from the air. It will dry very fast when thinly spread on paper. This property of rosin oil inevitably led to its being tried by makers of printing ink as a partial substitute for the dearer linseed oil. These attempts met with so much success that inks are now made with very little linseed in them indeed. Since the purification of rosin oil has been perfected the objectionable smell which was at one time an obstacle to its use can now be almost totally removed. Cheap inks, such as are used for newspapers, are now all made in this way.

Recipes.

A. Linseed Oil Vehicles.

I.

	Weak.	Medium.	Strong.
Much thickened linseed oil - - -	100	100	100
Rosin - - - - -	25	25	25
Rosin soap - - - - -			3
Thin linseed oil - - - - -	1–7	4–5	–

II.

	Weak.	Medium.	Strong.
Much thickened linseed oil - - -	100	100	100
Rosin - - - - -	50	50	50
Rosin soap - - - - -	10	10	10
Thin linseed oil - - - - -	8–9	6–7	0–1

III.

	Weak.	Medium.	Strong.
Much thickened linseed oil - - -	100	100	100
Rosin - - - - -	75	75	75
Rosin soap - - - - -	$7\frac{1}{2}$	$7\frac{1}{2}$	$7\frac{1}{2}$
	10–12	8–9	$1\text{–}1\frac{1}{2}$

B. Linseed and Rosin Oil Vehicles.

I.

Weak Composition.

	Weak.	Medium.	Strong.
Much thickened linseed oil - - -	100	100	100
Rosin oil - - - - -	100	100	100
Rosin - - - - -	50	100	150
Rosin soap - - - - -	6	10	15
Thin linseed oil - - - - -	6–7	15–18	10–12

II.

Medium Composition.

Much thickened linseed oil - -	100	100	100
Rosin oil - - - - -	100	100	100
Rosin - - - - -	50	100	100
Rosin soap - - - - -	6	10	15
Thin linseed oil - - - - -	7–8	10–12	15–18

Strong Composition.

	Weak.	Medium.	Strong.
Much thickened linseed oil	100	100	100
Rosin oil	100	100	100
Rosin	50	100	150
Rosin soap	6	10	15
Thin linseed oil	4–5	2–3	–

C. Linseed and Rosin Oils Compositions with Turpentine.

I.

Weak Composition.

	Medium.	Strong.
Unboiled linseed oil	100	35
Rosin oil	200	100
Rosin soap	4	7
Rosin	–	100
Turpentine	4	2

II.

Medium Composition.

Unboiled linseed oil	100	100
Rosin oil	25	10
Rosin soap	1	1
Rosin	–	10
Turpentine	1	1

III.

Strong Composition.

Unboiled linseed oil	30	3
Rosin oil	25	10
Rosin soap	1	1
Rosin	–	10
Turpentine	1	–

Making Printing Inks.

The manufacture of printing inks involves two operations, mixing the pigment with the vehicle and making the mixture homogeneous.

When it is a question of mixing in mineral pigments which usually have a considerable specific gravity, the work is simple enough. The vehicle being in a vat the finely powdered pigment is poured into it in a thin stream with constant stirring to get rid of air bubbles, and to mix it as perfectly as is then possible.

The pasty mass formed then goes straight to the colour mill. It is a better plan, however, to effect the preliminary mixing itself by mechanical means rather than by hand. Such machines can be procured and their use much shortens the passage of the ink through the colour mill. The case is different when the pigment to be mixed in is lamp-black. The particles are very minute, and of low specific gravity and contain quantities of entangled air. When the mixing is stirred it will be found that there are several practical difficulties. The lamp-black floats obstinately on the surface of the vehicle, is scarcely wetted by it, and even after long stirring the mixture is anything but satisfactory, and there is great loss of lamp-black, which flies into the air in dust. The best procedure is to put the lamp-black into the mixing vessel first, and then to stir in enough vehicle to make a thick paste, to which the rest of the vehicle is then added.

Under any circumstances, however, the manufacture of large quantities is inconsistent with hand labour, and machines are used which mix well and prevent all loss of lamp-black. The amount of lamp-black required for a given quantity of vehicle depends upon the nature of the lamp-black, and the sort of ink to be made, and varies from 20 to 30 per cent. of the weight of the finished ink. For making ordinary newspaper and poster inks, common lamp-black is used, while for printing books the ink is made from purified lamp-black, as the common sort would show brown on the white paper. The very finest lamp-black is used for the highest class of letterpress and art inks. Then, too, a blue black is often preferred to a full deep black, for the sake of appearance, and this is usually imparted by means of the purest Prussian blue.

Mixing Machines.

The machinery used in printing ink manufacture may be divided into two groups—the preliminary mixers, and the mills for making the rough mixture homogeneous. The preliminary mixer ought, however, to save a good deal of the work of the mill, whereby a large amount of time is saved. The best sort of mixer to use depends a good deal upon the amount of power available. If there is plenty of power at command at a cheaper rate, such as water power, rapid work is not so important as when more expensive sources of power, such as steam engines are used.

The Rolling Cask.

If water power is at hand, this form of mixing machine is much to be recommended. It is represented in Fig. 64.

It consists of an ordinary barrel or large drum, *a*, through the ends of which runs the axle *e* e^1, which does not, however, pass through the centre of the ends, but on a slant. On turning the crank shaft *c*, each end of the drum will be lifted in turn, and the contents thrown about. The materials are put in through the opening at *g* which is then closely fastened up. One end of the drum is fitted with the tap *f* to draw up the finished mass. The axle of the drum is put into connection with the motor, and the cask is turned, but not too fast, but at about twenty turns a minute only. This is quite fast enough

FIG. 64.

to get good mixing, and at a greater rate the centrifugal force would drive the contents too much against the sides, so that it would be difficult to empty the drum. How long the turning has to be kept up to get a good mix depends, of course, on the size of the drum and the amount of stuff in it. The drum should not be more than two-thirds full. To ascertain the time required, a sample is taken from the drum every half hour, and the drum must be kept going, until a sample spread on glass laid upon white paper shows no visible drops of vehicle or lumps of pigment. The mass can then go to the colour mill.

As many industries require machinery for mixing powders together or for mixing solids with liquids, many machine makers have occupied themselves with such apparatus, and we now proceed to describe some of the types.

Mixing and Kneading Machines.

A very suitable mixing machine, shown in Figs. 65 and 66, is that of Quack.

Figs. 65 and 66.

The mixer in this machine is a cylinder with a hemispherical bottom. Two specially formed blades are rotated inside by a fly-wheel worked by hand, or, if power is used, by a belt and pulley. When the contents of the vessel have been thoroughly mixed by the blades, they are turned out by tilting the vessel. The two blades rotate in opposite directions, and effect a very perfect mixing in a comparatively short time. If the machine is used for mixing the lamp-black into printing-ink, it must be closed with a well-fitting lid while at work, or large quantities of lamp-black will escape as dust when the mass is thrown about by the rotating blades.

An excellent mixing and kneading machine, which has now won for itself much favour in factories, is that of Werner and Pfleiderer. The construction of the machine is such that the motion of hand-stirring with a spoon in semi-circular sweep can be imitated with the same precision as that with which it kneads a pasty mass. These machines were at first made chiefly for the use of bakers, and enabled them to turn out perfectly mixed dough in a much less time than by hand labour. Fig. 67 shows the essential feature of this machine,

FIG. 67.

namely, the specially-formed stirrer, which rotates in the trough. As it comes very close to the bottom of the trough, it seizes upon the mass put into it, and crushes and kneads it in a similar fashion to the hands. The mass passes over the stirrer, and falls back to the bottom of the trough, where it is again kneaded, and so on, uninterruptedly, as long as the machine is in movement. In machines intended for mixing fine powders with liquids, the trough is provided with a well-fitting lid, which can be entirely removed when the machine is to be cleaned.

In using this machine for mixing lamp-black with a vehicle, it is better to proceed somewhat differently than with other mixing machines. Instead of putting powder and liquid together into the trough, and then running the machine until the mixing is complete, the whole of the lamp-black should be put into the trough first, and then only just enough vehicle to make rather a stiff paste. The rest of the vehicle is added at intervals during the working of the machine. In this way a mixture may be produced in many cases quite as good as can be got with a roller colour-mill, such as is shown in Fig. 68, and a perfectly satisfactory printer's ink will be the result.

FIG. 68.

In a printing-ink factory care must be taken to have one preliminary mixer big enough to keep all the colour-mills going. When this is the case, both power and machinery are employed to the fullest advantage, and the maximum of output can be obtained for a given time.

PRINTING INKS OF SPECIAL COMPOSITION.

We have now given the most approved compositions for inks for books and art printing, and supplied instructions for making them in a way which will meet with every requirement. Nevertheless new recipes are continually appearing, many of them patented, which direct the use in printing-ink making of all sorts of substances, very different from those generally employed. It is perfectly reasonable to look upon these recipes with a certain amount of suspicion, and we regard it as quite unnecessary to fill whole pages of our book with them. We shall confine ourselves to a selection, from the host of recipes, of those which in our opinion will really give useful inks. Among the inks deserving attention are those made with oleic acid, or with tar oil, and these will now be briefly described.

Oleic Acid Inks.

Oleic acid or oleine is the liquid which is obtained in stearine candle factories by putting cakes of fatty acid, got from the crude fat, into a hydraulic press. Stearic acid and palmitic acids, which are crystalline solids at the ordinary temperature, remain behind, but the pressure drives out the liquid oleic acid at first mixed with them. It is a yellow oily liquid, and is also called oleine in commerce. A property of oleic acid, important for our purpose, is the ease with which it can be emulsified with solution of carbonate of soda. This permits of the oleic acid ink being easily cleaned off the types.

Rösl's Oleic Acid Ink.

Mix together 9 lbs. of Austrian turpentine, 10 lbs. of soft soap, 4 lbs. of oleic acid, and 4 lbs. of lamp-black (more if a very black ink is wanted) by the aid of heat, and pass the whole through an ordinary paint mill. The type is cleaned after the use of this ink with a 1 per cent. solution of washing soda.

Artus's Oleic Acid Ink.

Heat 6 lbs. of Venice turpentine with 3 lbs. of oleic acid. Then mix in thoroughly 8 lbs. of soft soap. Then mix in 5 lbs. of lamp-black, 4 lbs. of Prussian blue, 2 lbs. of oleic acid and 2 lbs. of water. This does not differ greatly from Rösl's ink. It must be remembered that the oleic acid added last will not dissolve in the water added with it, so that all the oleic acid might as well be added at the first. The Prussian blue must be in the finest possible powder.

Coal Tar Oil Inks.

When coal tar is distilled, a whole series of products of various degrees of consistency is obtained, from liquids like water to thick oils. These bodies are separated by taking advantage of the differences in their specific gravity, and those oils having specific gravities between 0·85 and 0·89 can be used for making printing-ink.

According to one recipe, an excellent vehicle is prepared by boiling 100 lbs. of linseed oil with 6 lbs. of litharge, until a cold sample draws out into fine threads. Then add 40 lbs. of rosin and 20 lbs. of coal tar oil. Boil down to the desired consistency, adding a little coal tar oil occasionally if the mass is too thick. The ink is made by mixing lamp-black with this vehicle in the usual way.

G. Thenius's Printing-Ink.

This is a coal tar oil ink, and is prepared from the following remarkable recipe. Boil 25 lbs. of linseed oil with 3 lbs. of litharge until a cold sample is fairly thick. Then add 10 lbs. of pale American rosin previously melted. When all the rosin is dissolved add 5 lbs. of coal tar oil. Then remove from the fire and stir till cold. Before the coal tar can be used, it must be purified by mixing every 100 lbs. of it with ½ lb. of bichromate of potash, ¼ lb. of manganese peroxide, and 2 lbs. of sulphuric acid, in a vessel lined with lead. After stirring up well, let the mixture stand for a few hours. Then draw off the clear liquid from the sediment, and wash the liquid till it is quite neutral, first with dilute caustic soda, and then with water. This treatment of the coal tar oil with powerful oxidising agents is probably intended to destroy the powerful odour which the crude oil possesses.

XVI. OTHER LAMP-BLACK INKS.

Besides the lamp-black coloured masses used for letterpress printing inks, there are some others used for different graphic purposes which differ materially in their composition from ordinary printing inks. These products include Indian ink, and lithographic inks or "chalks." As regards the first article it came originally from China where as well as in other Asiatic countries it is employed for the purpose of writing. In Europe it is chiefly used for engineering and architectural designs, and in the plans drawn for the sale of estates. Lithographic inks are liquids which can be used either for drawing or for writing, and differ in two respects from ordinary writing inks. In the first place they must resist the action of certain reagents, and in the second they must be entirely removable from the stone by proper solvents. The lithographic "chalk," erroneously so called, is a substance with which we can write on stone as with a pencil on paper and must have the two properties above-mentioned. Both in lithographic and in Chinese inks, the colouring matter is finely divided carbon, the preparation of which has already been alluded to over and over again. They can be prepared with any carbon pigment, vine black, yeast black, bone black, etc., etc., provided the pigment is sufficiently finely divided. The art of preparing lamp-black for special purposes, especially for Chinese ink and the finest art printing inks, is the trade secret of many manufacturers. Every one who has properly learnt the methods of lamp-black making, however, will see that there can be no secret about the matter, for every pure lamp-black is one and the same substance, viz., amorphous carbon.

In former times the idea was that every thing depended upon what raw material was used for the making of the lamp-black, but it is known that that is a secondary matter and that the great secret lies in freeing carbon from the impurities which are mixed with it when it is first made. Now the impurities depend upon the raw material. Those adhering to lamp-black made from rosin are different to those it contains when made from rosin oil, and these are again different from the contaminations inseparable from the preparation of lamp-black from petroleum, while the foreign bodies in raw lamp-black made from fats are different again. Hence a certain amount of experience will enable one to tell by smelling the crude carbon from what raw material it was prepared. Whatever the raw material may be the treatment above described with caustic soda removes all the impurities and the final result is then the same in all cases.

It will now be clear that every maker of lamp-black ought to choose a raw material containing the largest percentage of carbon in proportion to the price so as to give the largest yield of lamp-black and must take great care to free his lamp-black from impurities by a proper treatment with caustic soda. Another important matter in the manufacture of Indian ink and ordinary printing inks of good quality is the thorough amalgamation of the lamp-black with the vehicle. If this point is neglected, the very finest lamp-black will be wasted in the production of very inferior inks.

Manufacture of Indian Ink.

The lamp-black used for this purpose is commonly quoted as an example of a special kind of black. Indian ink came to us first from China where it is used for writing after having been rubbed up with water. For a long time after it was imported into Europe nothing was known of the method of manufacture. Since China has been more and more opened up, the process of manufacture has been discovered and it may be briefly described as follows:—

In a small brick stove connected with a pipe having a gentle upwards slope sesame oil is burnt in a primitive lamp consisting of an oil holder and a wick. The gases from the smoky flames pass up the inclined pipe. This pipe is lengthened by paper tubes stuck end into end until no smoke can get out but is all deposited in the pipes. When a certain amount of oil has been burnt the operation is suspended and the tubes furthest from the stove are first taken down, those namely which contain the finest particles of lamp-black. Only those very finest particles are used for making the ink. The rest of the production is coarser and is rendered more impure by tarry bodies. This is used for the manufacture of coarser and cheaper pigments.

As the unpleasant smell of acroleine, a decomposition product of all fats, adheres even to this very finest lamp-black, the Chinese manufacturers hide it by adding a little camphor and musk. The work in China is all hand labour. The lamp-black is mixed with gum solution to a thick paste and then rolled into sheets with wooden rollers. The sheets are folded up and rolled out again, again folded, and again rolled, and so on till the mass is perfectly uniform. It is then pressed into moulds, and dried in the air. Any cracks which appear in the pieces are stopped with some of the fresh paste. The fully dry sticks, which are rather hard, have hieroglyphics painted on them in gold for the best sorts, and are packed in tissue paper.

To ascertain whether the Chinese method really gives better Indian ink than that prepared in Europe from highly purified lamp-black we carried out the Chinese process exactly as above described, burning sesame oil and following every detail exactly. The finest of the lamp-black was moulded with the best gum arabic and tragacanth mucilage and scented with a little of the tinctures of musk and camphor. The rolling was done by hand exactly as described. The result was a product quite equal to the finest Chinese manufacture in every respect. Another lot of paste was made by mixing by machinery. Microscopic examination showed, as might have been expected, that the machine mixed sample was much more uniform than that mixed by hand after the Chinese fashion. Finally, a mass was prepared with lamp-black prepared from fish oil and carefully purified with caustic soda. The three sorts were then compared, that made in China, that made by us from sesame oil in the Chinese way, and that made from fish oil. No difference between the three inks could be detected. This shows that if a product is carefully made from purified lamp-black the possible limit of excellence has been reached.

It has been reported by various travellers who were afforded the opportunity of visiting Chinese Indian ink factories, that different raw materials are used for making the lamp-black. Oils, fats and the resins of various trees are all mentioned. As even in China itself these inks are met with in very different qualities we may take it for granted that oil or fat lamp-black is used for the finer sorts and the coarser resin lamp-black for the inferior kinds. Common Chinese inks show clearly the characteristic brownish mark produced by unpurified lamp-black. The art of purifying lamp-black by calcination or by treatment with caustic soda seems unknown to the Chinese.

The lamp-black is bolted through very fine silk cloth, boiled with glue water with constant stirring and then kneaded with the hands exactly like dough. When the mass is uniform it is heated to facilitate the shaping of it, and pressed into moulds where it is allowed to dry slowly.

Fine sorts of Chinese inks are always very carefully got up. The sticks generally bear a large number of Chinese characters painted in blue and red and are often wrapped in real gold leaf.

On attempting to make the inks by this Chinese method it will be found that the setting of the glue hinders the mechanical working of the mass. This can, however, be easily avoided by preventing the glue from gelatinising by the use of a little acetic acid. The small amount added evaporates during the drying of the ink, but to prevent it from attacking the metal moulds, the insides of them are lightly rubbed with fine wood oil.

Lithographic Inks and Chalks.

Lithography consists essentially in making a drawing with lithographic ink or chalk on a specially prepared close and fine grained limestone. A liquid is then poured on which dissolves the surface of the stone, except where it is protected by the ink. The stone is then rinsed free from the etching liquid and a new liquid poured on. This dissolves the ink, so that the places where it was remain in relief when the stone is cleaned with a sponge. If then a printing roller is passed over the stone, it leaves the ink on the raised parts from which paper can then be printed in a press.

There is a very large number of recipes for lithographic inks, differing, however, more in the relative proportions than in the nature of the ingredients. Most of these compositions contain wax, soap, and shellac, together with fine lamp-black as a pigment. Many of them contain spermaceti also, but this can be replaced by tallow without seriously deteriorating the quality of the ink.

In trying a new recipe always commence by using very small quantities. An ounce or so will show whether the recipe answers its purpose. If it does not it is as well to try the effect of varying proportions of the ingredients. When success has been achieved it will be time enough to start making a large quantity. The reason why several attempts have usually to be made before a satisfactory result can be obtained, even with an approved recipe, is that the ingredients are rarely of the same quality. The soap used may contain more water than that demanded in the recipe. The wax often shows very different properties according to the degree of bleaching it has undergone, and the shellac may have the wrong melting point.

Most recipes which really give good results direct that the ingredients should be heated strongly enough to catch fire, and should be allowed to burn for a short time. We have, however, convinced ourselves that this burning is nothing more or less than sheer waste of valuable material. It is sufficient to heat close up to the ignition point without actually reaching it.

Hence a metal pot with a well-fitting lid should always be used for the preparation of lithographic inks. The lid will at once extinguish the ink if it should catch fire by chance. When the flame has been put out the pot must be taken from the fire and allowed to cool a little before removing the lid or the ink may again take fire when that is done.

Recipes for Lithographic Inks.

1. *German Recipes.*

Bleached virgin wax -	80 lbs.	
White grain soap (tallow) -	20 ,,	as free from water
Ruby shellac - - -	20 ,,	as possible.
Finest flame lamp-black -	20–25 ,,	

The soap must be dried by shaving it in fine shreds. This is best done with a carpenter's plane, and the shavings are spread out on paper near a stove. As the soap loses its water, the shavings rapidly shrink and finally form a mass easily rubbed to powder. The dried soap must be kept from the air in well-closed vessels, as it soon absorbs water vapour if exposed to it.

Having first melted the wax in the pot, heat it till it begins to develop the characteristic smell and then gradually add the soap. An assistant continually stirs the mass all the time with a wide paddle. When the mixture heats up till white fumes begin to rise, this marks a close approach to the ignition point so that on the appearance of the fumes the pot is removed from the fire and the lamp-black is stirred in bit by bit. The pot is then replaced on the fire and heated till the mass is permanently fluid. The stirring must be kept up till a sample spread out thinly on a sheet of glass appears quite homogeneous. If the mixture is not homogeneous the final product will be wanting in uniformity and be blacker in some places than in others. The shellac is next scattered in fine powder over the hot mass which is vigorously stirred all the time. When a sample on glass appears uniform, allow the mass to cool till it is just liquid enough to be poured into moulds where it soon sets into hard shining masses.

The following six recipes are made up exactly as above directed. The quantities are pounds avoirdupois:—

2. *English Recipes.*

	1	2	3	4	5	6
Yellow wax	40	—	140	18	6	100
Mastic	10	60	—	24	6	—
Gum lac	22	—	100	—	—	—
Tallow soap	22	60	70	18	6	100
Lamp-black	9	50	32	15	3	50
Shellac	–	300	—	36	30	100
Soda ash	–	60	—	—	6	50
Tallow	–	—	—	18	—	50
Turpentine	–	—	—	3	—	—
Rosin	–	—	10	—	—	—

The chief difference between English and German recipes for lithographic inks is that the latter often contain fat, usually mutton suet. With this soda is used and forms an emulsion with the fat. In general English lithographic inks are rather softer than those made from German recipes.

Lithographic Chalks.

Chalks or crayons for lithography are closely similar in composition to the lithographic inks, and like them must be easily removable from the stone while resisting the action of the etching liquid. They must in addition be capable of being sharpened with a knife like a lead pencil. The mass for moulding is made by melting the ingredients together and stirring them well until the mixture is quite orm. It is then cast into moulds and when it has set to a certain nt it is rolled out into thin rods. We subjoin four recipes with tities in pounds avoirdupois.

1. *For Strong Etching.*

Wax	64	64
Oil soap	22	48
Tallow	24	—
Saltpetre	2	3
Lamp-black	15	24
Spermaceti	—	32
Shellac	—	16

2. *Hard Chalk.*

Wax	50	60
Shellac	35	25
Soap	35	45
Soda	5	5
Tallow	5	10
Lamp-black	20	20

XVII. INKS FOR TYPEWRITING MACHINES.

In the typewriter a steel type at the end of a lever strikes against a dyed silk ribbon and forces the part of the ribbon struck against the paper which is supported by an india-rubber roller so as to make an impression of the letter on the paper. To avoid changing the ribbons frequently they are made very long, twenty yards or more. Every blow of a lever rolls the ribbon up a little so that the next type strikes it in a fresh place. In later types of machine the broad ribbon has also a lateral movement so that the types do not hit all along one line on the ribbon which would soon wear it out. Hence the places struck from a wavy line on the ribbon, and the ribbon can be used for a long time as the blows never hit exactly the same part of it again.

It is easy to see that unless the ink used is one that goes a very long way it will soon give out in spite of all these ingenious contrivances. Each letter made uses some of it. Hence for dyeing typewriter ribbons a colour must always be chosen which is distinguished by a high degree of colouring power so that a slight blow of the smallest steel type is sufficient to make a distinct impression on the paper. Even the most powerful colour would, however, not last long if it simply lay upon the face of the ribbon.

The ribbon must store up considerable quantities of colouring matter in its pores which will replenish the surface when it is exhansted, so that as long as there is any dye in the ribbon it is distributed uniformly through the fabric. A properly prepared ribbon twenty-two yards long will suffice to write thousands of pages of forty lines of seventy letters each, before the dye is so far exhausted that the ribbon must be changed for a new one.

It will readily appear from the foregoing that in making good typewriter ribbons the attention must not be confined to the colouring matter. The textile fabric must possess sufficient porosity to serve as a reservoir for a considerable quantity of dyestuff.

TYPEWRITER RIBBONS.

Choose a thin but closely woven material of which both warp and weft must be silk. This fabric, which must have at least the ribbon length of twenty-two yards, is cut into strips a little narrower than the distance between the flanges of the reels which are to take the ribbon. In cutting the strips care must be taken to cut exactly parallel to the direction of the warp, or some of the warp threads will work loose when the machine is used, catch in the mechanism and affect the running at last to such an extent that the whole machine will require a thorough cleaning. In a properly cut ribbon no threads get loose even with long wear, and no harm is done to the machine. No silk fabric containing any sort of size or dressing whatever can be used for typewriter ribbons. Not only is a dressed or weighted silk too hard, but it is not sufficiently porous. To test a piece of silk wash it first in luke-warm water, and then with luke-warm soap and water rinse and dry stretched on a frame. When dry it must have exactly the same feel and appearance as at first.

THE INKS.

As already mentioned, inks for typewriters must go a long way and must be such as will distribute themselves through the silk in a permanently liquid form. There is only one sort of colouring matter which possesses the first quality to a sufficient degree, the coal-tar dyes, which surpass all other colouring matters to an incredible extent in strength. For this reason no other inks are used for typewriter ribbons. Those used are either blues or violets and several dyes can be used for these colours. A water soluble blue or methyl violet are those most commonly employed. But other coal-tar dyes can be used at will, whether they are soluble in water or not, for those insoluble in water may be dissolved in spirit. As, moreover, most coal-tar dyes are soluble in glycerine, which must always enter into the composition of a typewriter ink, the maker has a fairly free hand in his choice of a dye.

It is often desirable to get a black ink for typewriters, and such a one can be made with Nigrosine but it must not be forgotten that this dye although it may be very finely divided enters into no real solution. This property makes it very difficult to prepare a typing ink from it which will give clean impressions. If the ink is made with an Induline blue, the colour of the writing is so deep that it looks nearly black, so that these blues are very suitable for use when very dark letters are required.

To make the ink, weigh out enough of the dye to serve for a number of ribbons. One hundred grammes (about 3½ oz.) is enough for a great many ribbons. Then take the same weight of pure glycerine as of dye and dilute it with its own weight of water. The glycerine must be weighed in a tared vessel for the amount of glycerine used is estimated by a second weighing. The dye is put into a porcelain basin and moistened with about half the dilute glycerine. The basin is then gently heated, stirring the contents with a glass rod. Most coal-tar dyes are freely soluble in glycerine and the solution in the basin is promoted by the warming. If the dye is one which is insoluble in water and but sparely soluble in glycerine add a little strong spirit, but not while the basin is over a flame for fear of fire.

To ascertain when all the dye is dissolved see if any solid particles can be felt with the end of the glass rod and let a drop of the mass from the end of the rod into a glass of clean water. The drop must become a perfectly transparent thread of clear colour and sink to the bottom. On stirring the whole of the water must become uniformly coloured and no solid particles must be seen floating in it. When this test shows that complete solution has occurred stop heating and allow the matter to cool. In cooling the matter often becomes gritty. This shows that some of the dye has crystallised out because there is not enough glycerine to keep it in solution at the ordinary temperature. If this occurs add some more of the dilute glycerine and heat up again for as long as there is solid matter in the cold mass it is unfit for use. When the mass is no longer gritty when cold it is ready for the ribbons. Dye a sample piece of ribbon a few yards long in the mass and dry it in a stretched state. It should then yield dye to the finger when lightly pressed. The ribbon is then put on a typewriter and tested by actual use. If it gives weak and pale letters there is not enough glycerine in the ink. If, on the other hand, the impressions are broad and letters which enclose a space such as o, b, a, and e, have the space filled up with dye there is too much glycerine and therefore the dye smudges.

The function of the glycerine is to keep the ink liquid. As the ink dries on the ribbon, the water or spirit used as a solvent gradually evaporates, and the ink would get quite dry were it not for the fact that the glycerine does not evaporate but, on the contrary, absorbs water from the air. The ink, which is for the most part dissolved in the glycerine (a little may in time separate in small crystals), is kept by it distributed through the whole substance of the ribbon. When the type impinges on the ribbon, ink is put on to the paper from the surface of the ribbon, but is immediately replaced by ink from the interior of the silk. We see here the reason why many typewriter ribbons give much better impressions after having been in use for a time than they did when new. The ink has tc be made rather too fat, *i.e.*, with a small excess of glycerine. When by use some of the ink has been removed from the ribbon the ink in the pores of the fabric has again spread uniformly and has again acquired the proper consistency for giving very clear impressions.

Once the right composition of the ink has been hit upon by means of the above mentioned writing tests, a formula can be made for use on future occasions. As above stated the initial quantities were 100 grammes of coal tar dye, 100 of glycerine and 100 of water. If it is found that of the 200 grammes of liquid 156 have been used the formula will be as follows :—

Dye	100 grammes		
Glycerine	78 ,,		256
Water	78 ,,		

The inks when properly made are usually too thick to be easily used on the ribbons. All then that has to be done is to dilute them with water or spirit to the proper consistency. When the ribbon is dried the extra water or spirit evaporates, but the amount of dye remains the same and so does that of the glycerine. Hence the proper proportion between dye and glycerine does not change, and that is all that is necessary. Nevertheless the ink should not be diluted more than is absolutely necessary to get it into the ribbon, for an over diluted ink would adhere so weakly to the silk that extra dye would then have to be painted on to the ribbon, a very troublesome and lengthy process.

Inking the Ribbons.

The ribbons are not properly prepared until the ink is uniformly distributed over the surface and through the entire thickness. This result can be secured only by a special process. If the ribbons are stretched out and then painted with the ink with a brush, they are got so that they can be used certainly, but a comparison of several pages written in succession will show that the strength of the impression varies very much. This shows that there is too much ink in some parts of the ribbon and too little in others; whereby some of the writing is too heavy and some too light.

The Inking Apparatus.

The inking is best done by dipping the ribbons and then pressing them. There is for the purpose a very simple apparatus whereby the ribbons can be inked very uniformly and rapidly. It consists of a rectangular dish of glass or porcelain to hold the ink. Over one of the ends of the dish a glass rod is held horizontally by supports. Twelve to sixteen inches above the middle of the dish is a wringer consisting of two rollers of soft vulcanised rubber. One of the rollers has fixed bearings, but the other can be adjusted by means of springs so as to press with more or less force against the first. The ribbons are laid in the ink so that they can be lifted out without getting entangled. This is best done by putting them in layers one above another parallel to the length of the dish. A ribbon is left to soak for several hours. One end of it is then put between the rollers, and the pressure between is regulated by screwing up the springs till while the superfluous ink is squeezed out, the ribbon comes from between the rollers quite wet as if it had been drawn through water. The excess of ink falls back into the dish as the rollers turn and the ribbon passes between them. On leaving the rollers the ribbon is at once stretched out so that it cannot shrink on drying and is kept stretched for at least twelve hours. It is then as dry as the glycerine in it will permit and should be at once rolled up.

Many typewriters have two fixed reels and the ribbon is reeled on to one as it leaves the other. Ribbons intended for such a machine are rolled up on a wooden reel and then packed for dispatch. In machines of more recent construction the ribbon reels can be taken out, and the ribbons are sold on the reels which have to be used in the typewriter.

The finished ribbons must be packed in such a way that they are preserved from damage during transit and from the access of dust. A good plan is to wrap the rolled ribbon in paraffined paper with a stronger wrapper over that. The parcels thus made should be packed in metal boxes, and will then be safe from dust or damage.

Making Carbon Papers.

Modern typewriters are so made that copies can be taken at the same time that the original writing is being done. With good carbon papers as many as six copies can be done at once. The copying process is as follows:—Under the paper which is to receive the original is laid a carbon paper with the coloured side away from the first paper. Next to this coloured side is a second sheet of white paper. If the keys are struck with some force the carbon paper will mark the second paper exactly as the ribbon marks the first. If the carbon paper is very thin and the copying paper is thin enough it is possible to get six quite distinct copies. Even more can be made, but the extra ones are less distinct, and, moreover, in taking several copies much force has to be used on the keys of the machine and this is a severe tax on the endurance of the mechanism.

Carbon Paper.

A very careful choice of paper must be made to begin with. The sheets to be made into carbons must be soft and thin, and, at the same time, very strong. It is evident that this is a rare combination of qualities, and anyone who wishes to make carbon papers on a large scale must come to some agreement with the paper maker.

Hitherto the papers chiefly used for the purpose have been of a very long stapled material and without any sizing. Such papers answer the purpose fairly well. Of late, however, a paper has been made for this use far surpassing all others that we have had the opportunity of trying. It is made by steeping a thin, fine, long stapled silk paper in a solution of viscose. When thoroughly soaked the paper is removed and spread out flat. After lying like this for a time the paper assumes the appearance of a very thin gelatinised skin. Before it is quite dry it must have weights piled on it or be run between rollers to prevent the surface from corrugating. The viscose solution is made by putting fine and well cleaned cotton into a porcelain vessel and treating it with strong caustic soda lye. The cotton is well stirred and kneaded with a glass or porcelain pestle to ensure its thorough penetration by the lye. When this has been done the vessel is left in a cool place at a temperature not exceeding 5–6° C. for a few hours. At this temperature the cellulose of the cotton combines with soda forming soda-cellulose, which does not differ in appearance from ordinary cellulose. When it has been formed the excess of lye is squeezed out from it very thoroughly so that the mass hardly appears damp. It is then at once put into a glass vessel which can be tightly closed with a well fitting stopper. Here it is treated with a weight of bisulphide of carbon equal to about 10 per cent. of that of the original cotton. The stopper is firmly inserted and the vessel is put in a cool place.

Bisulphide of carbon is very volatile and inflammable, and should never be poured on to the cellulose by artificial light. After the bisulphide has been in contact with the soda cellulose for about twelve hours the mass turns yellow from the formation of viscose. The stopper is then removed, and when the excess of bisulphide has evaporated water is poured over the viscose. Soon after this has been done, the mass swells up greatly and dissolves to a thick sticky liquid, which, when exposed to the air, decomposes, reproducing pure cellulose. Hence by pouring it over a glass plate a perfectly transparent skin of pure cellulose can be obtained. When the silk paper is soaked in viscose this penetrates completely into the paper so that when it decomposes it fills all the pores of the paper with cellulose and gives it the leathery nature above alluded to, and makes it very strong, while it remains soft. thin, and supple, and thus possesses the very qualities required for a carbon paper.

Ink for Carbon Paper.

This ink is made exactly in the same way as that for typewriter ribbons, except that once the proportions between glycerine and dye have been hit off, no water is used, nothing but dye and glycerine.

The simplest method is to take a large porcelain basin, put the weighed glycerine in it and heat slowly to 10° C. The dye is then stirred into the hot glycerine and the stirring is continued until solution is complete. The finished ink must be thick, but on no account gritty from undissolved dye. If this should, however, be the case, a little more glycerine will put matters right. Many makers add a little fatty oil to the ink, but the quantity must be very small or the impression from the paper will probably show an unsightly greasy border. Three to four per cent. of the total is the extreme limit allowable, and the oil is preferably castor oil, an oil which dissolves readily in spirit, whereby its through amalgamation with the ink is made easy. The dye having been stirred into the glycerine, the castor oil is added last in solution in spirit, and the whole mass is worked till quite homogeneous. During the rubbing most of the spirit evaporates and a very brief heating of the finished ink will free it from all spirituous smell.

The ink is spread as uniformly as possible with a wide brush over a level plate of stone or glass, and rolled out uniformly with a printing ink roller so that a rather thick coating of the ink remains on the roller. The sheet of paper to be inked is also laid on a level plate under a sheet of strong paper in which a four-cornered hole is cut a little smaller than the paper to be inked. The object of this upper sheet is to prevent any ink getting on the under side of the sheet to be inked. It is important that the ink should only be one side, for otherwise the paper would make impressions where they were not wanted. The inked roller is then passed over the sheet to be inked, which becomes penetrated with colour to a certain depth. The ink must, however, not be thin enough to get right through. The roller is then again inked and passed over the paper in a direction at right angles to the first to get the distribution of ink as uniform as possible. The rolling is continued, re-inking the roller each time, until the ink on the paper is thick enough to appear as a uniform shining layer which can for the most part be shaved off the paper with a thin knife blade. No ink must stick to a finger passed lightly over the paper. The finished carbon papers should be kept in a well closed box to prevent them from getting too dry, and in the box should be interleaved with ordinary writing paper.

When a copy has been taken the mark of every letter can be plainly seen on the carbon paper by looking obliquely at the inked surface. The parts where the types have struck are dull instead of shining, having lost a large part of their ink. If the carbon is to be used a second time it is put into the typewriter so that the types will strike between the lines formed on the paper the first time. In this way the sheet of carbon can be used again and again until its whole surface has been brought into requisition.

Rubber Stamp Inks.

The ink used on vulcanised rubber stamps should be such that when applied to a suitable pad it remains sufficiently fluid to adhere to the stamp. At the same time the fluidity should cease by the time the stamp is pressed upon an absorbing surface such as paper.

Formerly these inks were made by rubbing up pigments in fat to a paste. Such inks can, however, hardly be prevented from making impressions surrounded by a greasy mark caused by the fat spreading in the pores of the paper. Now, however, most stamping inks are made without grease and a properly prepared stamping ink contains nothing but glycerine and coal tar dye. As nearly all these dyes dissolve in hot glycerine the process of manufacture is simple enough. The dye, fuchsine, methyl violet, water blue, emerald green, etc., is put into a thin porcelain dish over which concentrated glycerine is poured, and the whole is heated to nearly 100° C. with constant stirring. It is important to use no more glycerine than is necessary to keep the dye dissolved when the ink is cold. If the mass turns gritty on cooling it must be heated up with more glycerine till solution is perfect.

In dealing with coal tar dyes insoluble in glycerine, or nearly so, dissolve them first in the least possible quantity of strong, hot alcohol. Then add the glycerine and heat till the spirit is evaporated.

To see whether the ink is properly made spread some of it on a strip of cloth and try it with a rubber stamp. On paper, the separate letters must be quite sharp and distinct. If they run at the edges there is too much glycerine in the ink and more dye must be added to it. If, on the contrary, the impression is indistinct and weak, the ink is too thick and must be diluted by carefully adding glycerine.

Indelible Stamping Inks.

All stamp inks made with coal tar dyes can often be removed by washing with spirit or with lye. The so-called indelible marking inks made with nitrate of silver can be absolutely removed from the fabric with cyanide of potassium solution.

For many trade purposes, *e.g.*, when a manufacturer sends his goods to be bleached or dyed, it seems to be of importance to stamp them with an ink which cannot be removed by washing or by treatment with any chemical. A stamp ink answering to this description is nigrosine, a black compound which cannot be destroyed in a fabric. But to make nigrosine adhere to the fabric in really indestructible fashion it must be developed in the fibre. To do this cut a stencil in strong cardboard, making a hole in the middle just large enough to receive the stamp. Then make a solution of 4 oz. of sulphate of copper in a mixture of 20 oz. of water and 2 of glycerine. Paint the solution on to the fabric through the hole in the stencil and let it dry on. In the meantime prepare a solution of aniline salt in distilled water and thicken it with gum until it will give a sharp impression with the stamp. If the stamp wetted with the aniline solution is pressed on the coppered part of the fabric nigrosine is generated in the fibre and the impression is absolutely indelible.

Another equally indelible stamping ink is made by thoroughly rubbing up a concentrated solution of viscose with fine lamp-black or any other finely powdered pigment. When this ink is used with a stamp the decomposition of the viscose forms cellulose which encloses the colouring matter and fixes it to the fabric, and the cellulose envelope cannot be destroyed without also destroying that part of the fabric to which it adheres.

Stamp Ink for Glass.

This special stamp ink is an etching ink and is used to produce indestructible marks or writing on glass by chemical means. The ink developes hydrofluoric acid which corrodes the glass. To make it, prepare separately a solution of 30 oz. of sodium fluoride and 7 of potassium sulphate in 500 of water and one of 14 oz. of solid zinc chloride and 65 of concentrated hydrochloric acid in 500 of water. The two solutions are then mixed. If a goose quill is dipped in the mixture it will write on glass, and a dull etched place appears wherever it touches. The mixture can also be used with a rubber stamp. When the ink has been applied it is left on for a few hours to bite deep enough and is then rinsed off with water. Rubber stamps used with this ink must be thoroughly cleaned with weak lye directly afterwards as the are rapidly destroyed by prolonged contact with the ink.

Abridged Catalogue
OF
Special Technical Books.

INDEX TO SUBJECTS.

PUBLISHED BY

SCOTT, GREENWOOD & SO

8 BROADWAY, LUDGATE, LONDON, E.C.

(Paints, Colours, Pigments and Printing Inks.)

THE CHEMISTRY OF PIGMENTS. By ERNEST J. PARRY, B.Sc. (Lond.), F.I.C., F.C.S., and J. H. COSTE, F.I.C., F.C.S. Demy 8vo. Five Illustrations. 285 pp. Price 10s. 6d. net. (Post free, 10s. 10d. home; 11s. 3d. abroad.)

THE MANUFACTURE OF PAINT. A Practical Handbook for Paint Manufacturers, Merchants and Painters. By J. CRUICKSHANK SMITH, B.Sc. Demy 8vo. 200 pp. Sixty Illustrations and One Large Diagram. Price 7s. 6d. net. (Post free, 7s. 10d. home; 8s. abroad.)

DICTIONARY OF CHEMICALS AND RAW PRODUCTS USED IN THE MANUFACTURE OF PAINTS, COLOURS, VARNISHES AND ALLIED PREPARATIONS. By GEORGE H. HURST, F.C.S. Demy 8vo. 380 pp. Price 7s. 6d. net. (Post free, 8s. home; 8s. 6d. abroad.)

THE MANUFACTURE OF LAKE PIGMENTS FROM ARTIFICIAL COLOURS. By FRANCIS H. JENNISON, F.I.C., F.C.S. **Sixteen Coloured Plates, showing Specimens of Eighty-nine Colours, specially prepared from the Recipes given in the Book.** 136 pp. Demy 8vo. Price 7s. 6d. net. (Post free, 7s. 10d. home; 8s. abroad.)

THE MANUFACTURE OF MINERAL AND LAKE PIGMENTS. Containing Directions for the Manufacture of all Artificial, Artists and Painters' Colours, Enamel, Soot and Metallic Pigments. A text-book for Manufacturers, Merchants, Artists and Painters. By Dr. JOSEF BERSCH. Translated by A. C. WRIGHT, M.A. (Oxon.), B.Sc. (Lond.). Forty-three Illustrations. 476 pp. Demy 8vo. Price 12s. 6d. net. (Post free, 13s. home; 13s. 6d. abroad.)

RECIPES FOR THE COLOUR, PAINT, VARNISH, OIL, SOAP AND DRYSALTERY TRADES. Compiled by AN ANALYTICAL CHEMIST. 330 pp. Second Revised Edition. Demy 8vo. Price 10s. 6d. net. (Post free, 11s. home; 11s. 3d. abroad.) [*Just published.*

OILMEN'S SUNDRIES AND HOW TO MAKE THEM. Being a Collection of Practical Recipes for Boot Polishes, Blues, Metal Polishes, Disinfectants, etc., compiled from "Oils, Colours and Drysalteries". Crown 8vo. 130 pages. Price 2s. 6d. net. (Post free, 2s. 9d. home; 2s. 10d. abroad.)

OIL COLOURS AND PRINTERS' INKS. By LOUIS EDGAR ANDÉS. Translated from the German. 215 pp. Crown 8vo. 56 Illustrations. Price 5s. net. (Post free, 5s. 4d. home; 5s. 6d. abroad.)

MODERN PRINTING INKS. A Practical Handbook for Printing Ink Manufacturers and Printers. By ALFRED SEYMOUR. Demy 8vo. Six Illustrations. 90 pages. Price 5s. net. (Post free, 5s. 4d. home; 5s. 6d. abroad.)

THREE HUNDRED SHADES AND HOW TO MIX THEM. For Architects, Painters and Decorators. By A. DESAINT, Artistic Interior Decorator of Paris. The book contains 100 folio Plates, measuring 12 in. by 7 in., each Plate containing specimens of three artistic shades. These shades are all numbered, and their composition and particulars for mixing are fully given at the beginning of the book. Each Plate is interleaved with grease-proof paper, and the volume is very artistically bound in art and linen with the Shield of the Painters' Guild impressed on the cover in gold and silver. Price 21s. net. (Post free, 21s. 6d. home; 22s. 6d. abroad.)

HOUSE DECORATING AND PAINTING. By W. NORMAN BROWN. Eighty-eight Illustrations. 150 pp. Crown 8vo. Price 3s. 6d. net. (Post free, 3s. 9d. home and abroad.)

A HISTORY OF DECORATIVE ART. By W. NORMAN BROWN. Thirty-nine Illustrations. 96 pp. Crown 8vo. Price 1s. net. (Post free, 1s. 3d. home and abroad.)

WORKSHOP WRINKLES for Decorators, Painters, Paperhangers, and Others. By W. N. BROWN. Crown 8vo. 128 pp. Second Edition. Price 2s. 6d. net. (Post free, 2s. 9d. home; 2s. 10d. abroad.)

CASEIN. By ROBERT SCHERER. Translated from the German by CHAS. SALTER. Demy 8vo. Illustrated. Second Revised English Edition. 160 pp. Price 7s. 6d. net. (Post free, 7s. 10d. home; 8s. abroad.)

SIMPLE METHODS FOR TESTING PAINTERS' MATERIALS. By A. C. WRIGHT, M.A. (Oxon.)., B.Sc. (Lond.). Crown 8vo. 160 pp. Price 5s. net. (Post free, 5s. 3d. home; 5s. 6d. abroad.)

IRON-CORROSION, ANTI-FOULING AND ANTI-CORROSIVE PAINTS. Translated from the German of LOUIS EDGAR ANDÉS. Sixty-two Illustrations. 275 pp. Demy 8vo. Price 10s. 6d. net. (Post free, 10s. 10d. home; 11s. 3d. abroad.)

THE TESTING AND VALUATION OF RAW MATERIALS USED IN PAINT AND COLOUR MANUFACTURE. By M. W. JONES, F.C.S. A Book for the Laboratories of Colour Works. 88 pp. Crown 8vo. Price 5s. net. (Post free, 5s. 3d. home and abroad.)

For contents of these books, see List I.

THE MANUFACTURE AND COMPARATIVE MERITS OF WHITE LEAD AND ZINC WHITE PAINTS. By G. PETIT, Civil Engineer, etc. Translated from the French. Crown 8vo. 100 pp. Price 4s. net. (Post free, 4s. 3d. home ; 4s. 4d. abroad.)

STUDENTS' HANDBOOK OF PAINTS, COLOURS, OILS AND VARNISHES. By JOHN FURNELL. Crown 8vo. 12 Illustrations. 96 pp. Price 2s. 6d. net. (Post free, 2s. 9d. home and abroad.)

PREPARATION AND USES OF WHITE ZINC PAINTS. Translated from the French of P. FLEURY. 250 Pages Crown 8vo. [*In the press.*

(Varnishes and Drying Oils.)

THE MANUFACTURE OF VARNISHES AND KINDRED INDUSTRIES. By J. GEDDES MCINTOSH. Second, greatly enlarged, English Edition, in three Volumes, based on and including the work of Ach. Livache.

VOLUME I.—**OIL CRUSHING, REFINING AND BOILING, THE MANUFACTURE OF LINOLEUM, PRINTING AND LITHOGRAPHIC INKS, AND INDIA-RUBBER SUBSTITUTES.** Demy 8vo. 150 pp. 29 Illustrations. Price 7s. 6d. net. (Post free, 7s. 10d. home ; 8s. abroad.)

VOLUME II.—**VARNISH MATERIALS AND OIL-VARNISH MAKING.** Demy 8vo. 70 Illustrations. 220 pp. Price 10s. 6d. net. (Post free, 10s. 10d. home ; 11s. 3d. abroad.)

VOLUME III.—**SPIRIT VARNISHES AND SPIRIT VARNISH MATERIALS.** Demy 8vo. Illustrated. 464 pp. Price 12s. 6d. net. (Post free, 13s. home ; 13s. 6d. abroad.)

DRYING OILS, BOILED OIL AND SOLID AND LIQUID DRIERS. By L. E. ANDÉS. Expressly Written for this Series of Special Technical Books, and the Publishers hold the Copyright for English and Foreign Editions. Forty-two Illustrations. 342 pp. Demy 8vo. Price 12s. 6d. net. (Post free, 13s. home ; 13s. 3d. abroad.)

(*Analysis of Resins, see page* 9.)

(Oils, Fats, Waxes, Greases, Petroleum.)

LUBRICATING OILS, FATS AND GREASES: Their Origin, Preparation, Properties, Uses and Analyses. A Handbook for Oil Manufacturers, Refiners and Merchants, and the Oil and Fat Industry in General. By GEORGE H. HURST, F.C.S. Third Revised and Enlarged Edition. Seventy-four Illustrations. 384 pp. Demy 8vo. Price 10s. 6d. net. (Post free, 11s. home; 11s. 3d. abroad.)

TECHNOLOGY OF PETROLEUM: Oil Fields of the World—Their History, Geography and Geology—Annual Production and Development—Oil-well Drilling—Transport. By HENRY NEUBERGER and HENRY NOALHAT. Translated from the French by J. G. McINTOSH. 550 pp. 153 Illustrations. 26 Plates. Super Royal 8vo. Price 21s. net. (Post free, 21s. 9d. home; 23s. 6d. abroad.)

MINERAL WAXES: Their Preparation and Uses. By RUDOLF GREGORIUS. Translated from the German. Crown 8vo. 250 pp. 32 Illustrations. Price 6s. net. (Post free, 6s. 4d. home; 6s. 6d. abroad.)

THE PRACTICAL COMPOUNDING OF OILS, TALLOW AND GREASE FOR LUBRICATION, ETC. By AN EXPERT OIL REFINER. Second Edition. 100 pp. Demy 8vo. Price 7s. 6d. net. (Post free, 7s. 10d. home; 8s. abroad.)

THE MANUFACTURE OF LUBRICANTS, SHOE POLISHES AND LEATHER DRESSINGS. By RICHARD BRUNNER. Translated from the Sixth German Edition by CHAS. SALTER. 10 Illustrations. Crown 8vo. 170 pp. Price 7s. 6d. net. (Post free, 7s. 10d. home; 8s. abroad.)

THE OIL MERCHANTS' MANUAL AND OIL TRADE READY RECKONER. Compiled by FRANK F. SHERRIFF. Second Edition Revised and Enlarged. Demy 8vo. 214 pp. With Two Sheets of Tables. Price 7s. 6d. net. (Post free, 7s. 10d. home; 8s. 3d. abroad.)

ANIMAL FATS AND OILS: Their Practical Production Purification and Uses for a great Variety of Purposes. Their Properties, Falsification and Examination. Translated from the German of LOUIS EDGAR ANDÉS. Sixty-two Illustrations. 240 pp. Second Edition, Revised and Enlarged. Demy 8vo. Price 10s. 6d. net. (Post free, 10s. 10d. home; 11s. 3d. abroad.)

For contents of these books, see List I.

VEGETABLE FATS AND OILS: Their Practical Preparation, Purification and Employment for Various Purposes, their Properties, Adulteration and Examination. Translated from the German of LOUIS EDGAR ANDÉS. Ninety-four Illustrations. 340 pp. Second Edition. Demy 8vo. Price 10s. 6d. net. (Post free, 11s. home; 11s. 6d. abroad.)

EDIBLE FATS AND OILS: Their Composition, Manufacture and Analysis. By W. H. SIMMONS, B.Sc. (Lond.), and C. A. MITCHELL, B.A. (Oxon.). Demy 8vo. 150 pp. Price 7s. 6d. net. (Post free, 7s. 9d. home; 8s. abroad.)

(Essential Oils and Perfumes.)

THE CHEMISTRY OF ESSENTIAL OILS AND ARTIFICIAL PERFUMES. By ERNEST J. PARRY, B.Sc. (Lond.), F.I.C., F.C.S. Second Edition, Revised and Enlarged. 552 pp. 20 Illustrations. Demy 8vo. Price 12s. 6d. net. (Post free, 13s. home; 13s. 6d. abroad.)

(Soap Manufacture.)

SOAPS. A Practical Manual of the Manufacture of Domestic, Toilet and other Soaps. By GEORGE H. HURST, F.C.S. 2nd edition. 390 pp. 66 Illustrations. Demy 8vo. Price 12s. 6d. net. (Post free, 13s. home; 13s. 6d. abroad.)

TEXTILE SOAPS AND OILS. Handbook on the Preparation, Properties and Analysis of the Soaps and Oils used in Textile Manufacturing, Dyeing and Printing. By GEORGE H. HURST, F.C.S. Crown 8vo. 195 pp. 1904. Price 5s. net. (Post free, 5s. 4d. home; 5s. 6d. abroad.)

THE HANDBOOK OF SOAP MANUFACTURE. By WM. H. SIMMONS, B.Sc. (Lond.), F.C.S. and H. A. APPLETON. Demy 8vo. 160 pp. 27 Illustrations. Price 8s. 6d. net. (Post free, 8s. 10d. home; 9s. abroad.)

MANUAL OF TOILET SOAPMAKING, including Medicated Soaps Stain-removing Soaps, Metal Polishing Soaps, Soap Powders and Detergents. Translated from the German of Dr. C. Deite. Demy quarto. 150 pages. 79 Illustrations. Price 12s. 6d. net. (Post free, 13s. home; 13s. 6d. abroad)

(Cosmetical Preparations.)

COSMETICS: MANUFACTURE, EMPLOYMENT AND TESTING OF ALL COSMETIC MATERIALS AND COSMETIC SPECIALITIES. Translated from the German of Dr. THEODOR KOLLER. Crown 8vo. 2 2 pp. Price 5s. net. (Post free, 5s. 4d. home; 5s. 6d. abroad.)

(Glue, Bone Products and Manures.)

GLUE AND GLUE TESTING. By SAMUEL RIDEAL, D.Sc. (Lond.), F.I.C. Fourteen Engravings. 144 pp. Demy 8vo. Price 10s. 6d. net. (Post free, 10s. 10d. home ; 11s. abroad.)

BONE PRODUCTS AND MANURES: An Account of the most recent Improvements in the Manufacture of Fat, Glue, Animal Charcoal, Size, Gelatine and Manures. By THOMAS LAMBERT, Technical and Consulting Chemist. Illustrated by Twenty-one Plans and Diagrams. 162 pp. Demy 8vo. Price 7s. 6d. net. (Post free, 7s. 10d. home ; 8s. abroad.)

(*See also Chemical Manures, p.* 9.)

(Chemicals, Waste Products, etc.)

REISSUE OF **CHEMICAL ESSAYS OF C. W. SCHEELE.** First Published in English in 1786. Translated from the Academy of Sciences at Stockholm, with Additions. 300 pp. Demy 8vo. Price 5s. net. (Post free, 5s. 6d. home ; 5s. 9d. abroad.)

THE MANUFACTURE OF ALUM AND THE SULPHATES AND OTHER SALTS OF ALUMINA AND IRON. Their Uses and Applications as Mordants in Dyeing and Calico Printing, and their other Applications in the Arts Manufactures, Sanitary Engineering, Agriculture and Horticulture. Translated from the French of LUCIEN GESCHWIND. 195 Illustrations. 400 pp. Royal 8vo. Price 12s. 6d. net. (Post free, 13s. home ; 13s. 6d. abroad.)

AMMONIA AND ITS COMPOUNDS: Their Manufacture and Uses. By CAMILLE VINCENT, Professor at the Central School of Arts and Manufactures, Paris. Translated from the French by M. J. SALTER. Royal 8vo. 114 pp. Thirty-two Illustrations. Price 5s. net. (Post free, 5s. 4d. home ; 5s. 6d. abroad.)

CHEMICAL WORKS: Their Design, Erection, and Equipment. By S. S. DYSON and S. S. CLARKSON. Royal 8vo. 220 pp. With Plates and Illustrations. Price 21s. net. (Post free, 21s. 6d. home ; 22s. abroad.) [*Just published.*

MANUAL OF CHEMICAL ANALYSIS, as applied to the Assay of Fuels, Ores, Metals, Alloys, Salts and other Mineral Products. By E. PROST, D.Sc. Translated by J. CRUICKSHANK SMITH, B.Sc. Royal 8vo. 300 pages. 44 Illustrations. Price 12s. 6d. net. (Post free, 13s. home ; 13s. 6d. abroad.)

TESTING OF CHEMICAL REAGENTS FOR PURITY. Translated from the German of Dr. C. KRAUCH. Royal 8vo. 350 pages. Price 12s 6d. net. (Post free, 13s. home ; 13s. 6d. abroad.)

For contents of these books, see List I.

SHALE TAR DISTILLATION: The Treatment of Shale and Lignite Products. Translated from the German of W. SCHEITHAUER. [*In the Press.*

INDUSTRIAL ALCOHOL. A Practical Manual on the Production and Use of Alcohol for Industrial Purposes and for Use as a Heating Agent, as an Illuminant and as a Source of Motive Power. By J. G. McINTOSH, Lecturer on Manufacture and Applications of Industrial Alcohol at The Polytechnic, Regent Street, London. Demy 8vo. 1907. 250 pp. With 75 Illustrations and 25 Tables. Price 7s. 6d. net. (Post free, 7s. 9d. home; 8s. abroad.)

THE UTILISATION OF WASTE PRODUCTS. A Treatise on the Rational Utilisation, Recovery and Treatment of Waste Products of all kinds. By Dr. THEODOR KOLLER. Translated from the Second Revised German Edition. Twenty-two Illustrations. Demy 8vo. 280 pp. Price 7s. 6d. net. (Post free, 7s. 10d. home; 8s. 3d. abroad.)

ANALYSIS OF RESINS AND BALSAMS. Translated from the German of Dr. KARL DIETERICH. Demy 8vo. 340 pp. Price 7s. 6d. net. (Post free, 7s. 10d. home; 8s. 3d. abroad.)

DISTILLATION OF RESINS, RESINATE LAKES AND PIGMENTS, CARBON PIGMENTS AND PIGMENTS FOR TYPEWRITING MACHINES, MANIFOLDERS, ETC. By VICTOR SCHWEIZER. Demy 8vo. 185 pages. 68 Illustrations. Price 7s. 6d. net. (Post free, 8s. home; 8s. 3d. abroad.)

(Agricultural Chemistry and Manures.)

MANUAL OF AGRICULTURAL CHEMISTRY. By HERBERT INGLE, F.I.C., Late Lecturer on Agricultural Chemistry, the Leeds University; Lecturer in the Victoria University. Second Edition, with additional matter relating to Tropical Agriculture, etc. 438 pp. 11 Illustrations. Demy 8vo. Price 7s. 6d. net. (Post free, 8s. home; 8s. 6d. abroad.)

CHEMICAL MANURES. Translated from the French of J. FRITSCH. Demy 8vo. Illustrated. 340 pp. Price 10s. 6d. net. (Post free, 11s. home; 11s. 6d. abroad.)

(*See also Bone Products and Manures, p.* 8.)

(Writing Inks and Sealing Waxes.)

INK MANUFACTURE: Including Writing, Copying, Lithographic, Marking, Stamping, and Laundry Inks. By SIGMUND LEHNER. Three Illustrations. Crown 8vo. 162 pp. Translated from the German of the Fifth Edition. Price 5s. net. (Post free, 5s. 3d. home; 5s. 6d. abroad.)

SEALING - WAXES, WAFERS AND OTHER ADHESIVES FOR THE HOUSEHOLD, OFFICE, WORKSHOP AND FACTORY. By H. C. STANDAGE. Crown 8vo. 96 pp. Price 5s. net. (Post free, 5s. 3d. home; 5s. 4d. abroad.)

(Lead Ores and Lead Compounds.)

LEAD AND ITS COMPOUNDS. By THOS. LAMBERT, Technical and Consulting Chemist. Demy 8vo. 226 pp. Forty Illustrations. Price 7s. 6d. net. (Post free, 7s. 10d. home; 8s. 3d. abroad.)

NOTES ON LEAD ORES: Their Distribution and Properties. By JAS. FAIRIE, F.G.S. Crown 8vo. 64 pages. Price 1s. net. (Post free, 1s. 3d. home; 1s. 4d. abroad.)

(*White Lead and Zinc White Paints, see p.* 5.)

(Industrial Hygiene.)

THE RISKS AND DANGERS TO HEALTH OF VARIOUS OCCUPATIONS AND THEIR PREVENTION. By LEONARD A. PARRY, M.D., B.Sc. (Lond.). 196 pp. Demy 8vo. Price 7s. 6d. net. (Post free, 7s. 10d. home; 8s. abroad.)

(Industrial Uses of Air, Steam and Water.)

DRYING BY MEANS OF AIR AND STEAM. Explanations, Formulæ, and Tables for Use in Practice. Translated from the German of E. HAUSBRAND. Two folding Diagrams and Thirteen Tables. Crown 8vo. 72 pp. Price 5s. net. (Post free, 5s. 3d. home; 5s. 6d. abroad.)

(*See also "Evaporating, Condensing and Cooling Apparatus," p.* 19.)

PURE AIR, OZONE AND WATER. A Practical Treatise of their Utilisation and Value in Oil, Grease, Soap, Paint, Glue and other Industries. By W. B. COWELL. Twelve Illustrations. Crown 8vo. 85 pp. Price 5s. net. (Post free, 5s. 3d. home; 5s. 6d. abroad.)

THE INDUSTRIAL USES OF WATER. COMPOSITION—EFFECTS—TROUBLES—REMEDIES—RESIDUARY WATERS—PURIFICATION—ANALYSIS. By H. DE LA COUX. Royal 8vo. Translated from the French and Revised by ARTHUR MORRIS. 364 pp. 135 Illustrations. Price 10s. 6d. net. (Post free, 11s. home; 11s. 6d. abroad.)

(*See Books on Smoke Prevention, Engineering and Metallurgy, p.* 19, *etc.*)

For contents of these books, see List III.

(X Rays.)

PRACTICAL X RAY WORK. By FRANK T. ADDYMAN, B.Sc. (Lond.), F.I.C., Member of the Roentgen Society of London; Radiographer to St. George's Hospital; Demonstrator of Physics and Chemistry, and Teacher of Radiography in St. George's Hospital Medical School. Demy 8vo. Twelve Plates from Photographs of X Ray Work. Fifty-two Illustrations. 200 pp. Price 10s. 6d. net. (Post free, 10s. 10d. home; 11s. 3d. abroad.)

(India-Rubber and Gutta Percha.)

INDIA-RUBBER AND GUTTA PERCHA. Second English Edition, Revised and Enlarged. Based on the French work of T. SEELIGMANN, G. LAMY TORRILHON and H. FALCONNET by JOHN GEDDES MCINTOSH. Royal 8vo. 100 Illustrations. 400 pages. Price 12s. 6d. net. (Post free, 13s. home; 13s. 6d. abroad.)

(Leather Trades.)

THE LEATHER WORKER'S MANUAL. Being a Compendium of Practical Recipes and Working Formulæ for Curriers, Bootmakers, Leather Dressers, Blacking Manufacturers, Saddlers, Fancy Leather Workers. By H. C. STANDAGE. Demy 8vo. 165 pp. Price 7s. 6d. net. (Post free, 7s. 10d. home; 8s. abroad.)

(*See also Manufacture of Shoe Polishes, Leather Dressings, etc., p.* 6.)

(Pottery, Bricks, Tiles, Glass, etc.)

MODERN BRICKMAKING. By ALFRED B. SEARLE, Royal 8vo. 440 pages. 260 Illustrations. Price 12s. 6d. net. (Post free, 13s. home; 13s. 6d. abroad.)

THE MANUAL OF PRACTICAL POTTING. Compiled by Experts, and Edited by CHAS. F. BINNS. Third Edition, Revised and Enlarged. 200 pp. Demy 8vo. Price 17s. 6d. net. (Post free, 17s. 10d. home; 18s. 3d. abroad.)

POTTERY DECORATING. A Description of all the Processes for Decorating Pottery and Porcelain. By R. HAINBACH. Translated from the German. Crown 8vo. 250 pp. Twenty-two Illustrations. Price 7s. 6d. net. (Post free, 7s. 10d. home; 8s. abroad.)

A TREATISE ON CERAMIC INDUSTRIES. A Complete Manual for Pottery, Tile, and Brick Manufacturers. By EMILE BOURRY. A Revised Translation from the French, with some Critical Notes by ALFRED B. SEARLE. Demy 8vo. 308 Illustrations. 460 pp. Price 12s. 6d. net. (Post free, 13s. home; 13s. 6d. abroad.)

ARCHITECTURAL POTTERY. Bricks, Tiles, Pipes, Enamelled Terra-cottas, Ordinary and Incrusted Quarries, Stoneware Mosaics, Faïences and Architectural Stoneware. By LEON LEFÊVRE. Translated from the French by K. H. BIRD, M.A., and W. MOORE BINNS. With Five Plates. 950 Illustrations in the Text, and numerous estimates. 500 pp., royal 8vo. Price 15s. net. (Post free, 15s. 6d. home; 16s. 6d. abroad.)

CERAMIC TECHNOLOGY: Being some Aspects of Technical Science as Applied to Pottery Manufacture. Edited by CHARLES F. BINNS. 100 pp. Demy 8vo. Price 12s. 6d. net. (Post free, 12s. 10d. home; 13s. abroad.)

THE ART OF RIVETING GLASS, CHINA AND EARTHENWARE. By J. HOWARTH. Second Edition. Paper Cover. Price 1s. net. (By post, home or abroad, 1s. 1d.)

NOTES ON POTTERY CLAYS. The Distribution, Properties, Uses and Analyses of Ball Clays, China Clays and China Stone. By JAS. FAIRIE, F.G.S. 132 pp. Crown 8vo. Price 3s. 6d. net. (Post free, 3s. 9d. home; 3s. 10d. abroad.)

HOW TO ANALYSE CLAY. By H. M. Ashby. Demy 8vo. 72 Pages. 20 Illustrations. Price 3s. 6d. net. (Post free, 3s. 9d. home; 3s. 10d. abroad.)

A Reissue of

THE HISTORY OF THE STAFFORDSHIRE POTTERIES; AND THE RISE AND PROGRESS OF THE MANUFACTURE OF POTTERY AND PORCELAIN. With References to Genuine Specimens, and Notices of Eminent Potters. By SIMEON SHAW. (Originally published in 1829.) 265 pp. Demy 8vo. Price 5s. net. (Post free, 5s. 4d. home; 5s. 9d. abroad.)

A Reissue of

THE CHEMISTRY OF THE SEVERAL NATURAL AND ARTIFICIAL HETEROGENEOUS COMPOUNDS USED IN MANUFACTURING PORCELAIN, GLASS AND POTTERY. By SIMEON SHAW. (Originally published in 1837.) 750 pp. Royal 8vo. Price 10s. net. (Post free, 10s. 6d. home; 12s. abroad.)

BRITISH POTTERY MARKS. By G. WOOLLISCROFT RHEAD. Demy 8vo. 310 pp. With over Twelve-hundred Illustrations of Marks. Price 7s. 6d. net. (Post free, 8s. home; 8s. 3d. abroad.)

For contents of these books, see List III.

(Glassware, Glass Staining and Painting.)

RECIPES FOR FLINT GLASS MAKING. By a British Glass Master and Mixer. Sixty Recipes. Being Leaves from the Mixing Book of several experts in the Flint Glass Trade, containing up-to-date recipes and valuable information as to Crystal, Demi-crystal and Coloured Glass in its many varieties. It contains the recipes for cheap metal suited to pressing, blowing, etc., as well as the most costly crystal and ruby. Second Edition. Crown 8vo. Price 10s. 6d. net. (Post free, 10s. 9d. home; 10s. 10d. abroad.)

A TREATISE ON THE ART OF GLASS PAINTING. Prefaced with a Review of Ancient Glass. By ERNEST R. SUFFLING. With One Coloured Plate and Thirty-seven Illustrations. Demy 8vo. 140 pp. Price 7s. 6d. net. (Post free, 7s. 10d. home; 8s. abroad.)

(Paper Making, Paper Dyeing, and Testing.)

THE DYEING OF PAPER PULP. A Practical Treatise for the use of Papermakers, Paperstainers, Students and others. By JULIUS ERFURT, Manager of a Paper Mill. Translated into English and Edited with Additions by JULIUS HÜBNER, F.C.S., Lecturer on Papermaking at the Manchester Municipal Technical School. With illustrations and **157 patterns of paper dyed in the pulp.** Royal 8vo, 180 pp. Price 15s. net. (Post free, 15s. 6d. home; 16s. 6d. abroad).

THE PAPER MILL CHEMIST. By HENRY P. STEVENS, M.A., Ph.D., F.I.C. Royal 12mo. 60 illustrations. 300 pp. Price 7s. 6d. net. (Post free, 7s. 9d. home; 7s. 10d. abroad.)

THE TREATMENT OF PAPER FOR SPECIAL PURPOSES. By L. E. ANDÉS. Translated from the German. Crown 8vo. 48 Illustrations 250 pp. Price 6s. net. (Post free, 6s. 4d. home; 6s. 6d. abroad.)

(Enamelling on Metal.)

ENAMELS AND ENAMELLING. For Enamel Makers, Workers in Gold and Silver, and Manufacturers of Objects of Art. By PAUL RANDAU. Translated from the German. With Sixteen Illustrations. Demy 8vo. 180 pp. Price 10s. 6d. net. (Post free, 10s. 10d. home; 11s. abroad.)

THE ART OF ENAMELLING ON METAL. By W. NORMAN BROWN. Twenty-eight Illustrations. Crown 8vo. 60 pp. Price 2s. 6d. net. (Post free, 2s. 9d. home and abroad.)

(Textile and Dyeing Subjects.)

THE FINISHING OF TEXTILE FABRICS (Woollen, Worsted, Union and other Cloths). By ROBERTS BEAUMONT, M.Sc., M.I Mech.E., Professor of Textile Industries, the University of Leeds; Author of "Colour in Woven Design"; "Woollen and Worsted Cloth Manufacture"; "Woven Fabrics at the World's Fair"; Vice-President of the Jury of Award at the Paris Exhibition, 1900; Inspector of Textile Institutes; Society of Arts Silver Medallist; Honorary Medallist of the City and Guilds of London Institute. With 150 Illustrations of Fibres, Yarns and Fabrics, also Sectional and other Drawings of Finishing Machinery Demy 8vo. 260 pp. Price 10s. 6d. net. (Post free, 10s. 10d. home; 11s. 3d. abroad.)

FIBRES USED IN TEXTILE AND ALLIED INDUSTRIES. By C. AINSWORTH MITCHELL, B.A. (Oxon.), F.I.C., and R. M. PRIDEAUX, F.I.C. With 66 Illustrations specially drawn direct from the Fibres. Demy 8vo. 200 pp. Price 7s. 6d. net. (Post free, 7s. 10d. home; 8s. abroad.)

DRESSINGS AND FINISHINGS FOR TEXTILE FABRICS AND THEIR APPLICATION. Description of all the Materials used in Dressing Textiles: Their Special Properties, the preparation of Dressings and their employment in Finishing Linen, Cotton, Woollen and Silk Fabrics. Fireproof and Waterproof Dressings, together with the principal machinery employed. Translated from the Third German Edition of FRIEDRICH POLLEYN. Demy 8vo. 280 pp. Sixty Illustrations. Price 7s. 6d. net. (Post free, 7s. 10d. home; 8s. abroad.)

THE CHEMICAL TECHNOLOGY OF TEXTILE FIBRES: Their Origin, Structure, Preparation, Washing, Bleaching, Dyeing, Printing and Dressing. By Dr. GEORG VON GEORGIEVICS. Translated from the German by CHARLES SALTER. 320 pp. Forty-seven Illustrations. Royal 8vo. Price 10s. 6d. net. (Post free, 11s. home; 11s. 3d. abroad.)

POWER-LOOM WEAVING AND YARN NUMBERING, According to Various Systems, with Conversion Tables. Translated from the German of ANTHON GRUNER. **With Twenty-six Diagrams in Colours.** 150 pp. Crown 8vo. Price 7s. 6d. net. (Post free, 7s. 9d. home; 8s. abroad.)

TEXTILE RAW MATERIALS AND THEIR CONVERSION INTO YARNS. (The Study of the Raw Materials and the Technology of the Spinning Process.) By JULIUS ZIPSER. Translated from German by CHARLES SALTER. 302 Illustrations. 500 pp. Demy 8vo. Price 10s. 6d. net. (Post free, 11s. home; 11s. 6d. abroad.)

For contents of these books, see List II.

GRAMMAR OF TEXTILE DESIGN. By H. NISBET, Weaving and Designing Master, Bolton Municipal Technical School. Demy 8vo. 280 pp. 490 Illustrations and Diagrams Price 6s. net. (Post free, 6s. 4d. home; 6s. 6d. abroad.)

ART NEEDLEWORK AND DESIGN. POINT LACE. A Manual of Applied Art for Secondary Schools and Continuation Classes. By M. E. WILKINSON. Oblong quarto. With 22 Plates. Bound in Art Linen. Price 3s. 6d. net. (Post free, 3s. 10d. home; 4s. abroad.)

HOME LACE-MAKING. A Handbook for Teachers and Pupils. By M. E. W. MILROY. Crown 8vo. 64 pp. With 3 Plates and 9 Diagrams. Price 1s. net. (Post free, 1s. 3d. home; 1s. 4d. abroad.)

THE CHEMISTRY OF HAT MANUFACTURING. Lectures delivered before the Hat Manufacturers' Association. By WATSON SMITH, F.C.S., F.I.C. Revised and Edited by ALBERT SHONK. Crown 8vo. 132 pp. 16 Illustrations. Price 7s. 6d. net. (Post free, 7s. 9d. home; 7s. 10d. abroad.)

THE TECHNICAL TESTING OF YARNS AND TEXTILE FABRICS. With Reference to Official Specifications. Translated from the German of Dr. J. HERZFELD. Second Edition. Sixty-nine Illustrations. 200 pp. Demy 8vo. Price 10s. 6d net. (Post free, 10s. 10d. home; 11s. abroad.)

DECORATIVE AND FANCY TEXTILE FABRICS. By R. T. LORD. For Manufacturers and Designers of Carpets, Damask, Dress and all Textile Fabrics. 200 pp. Demy 8vo. 132 Designs and Illustrations. Price 7s. 6d. net. (Post free, 7s. 10d. home; 8s. abroad.)

THEORY AND PRACTICE OF DAMASK WEAVING. By H. KINZER and K. WALTER. Royal 8vo. Eighteen Folding Plates. Six Illustrations. Translated from the German. 110 pp. Price 8s. 6d. net. (Post free, 9s. home; 9s. 6d. abroad.)

FAULTS IN THE MANUFACTURE OF WOOLLEN GOODS AND THEIR PREVENTION. By NICOLAS REISER. Translated from the Second German Edition. Crown 8vo. Sixty-three Illustrations. 170 pp. Price 5s. net (Post free, 5s. 4d. home; 5s. 6d. abroad.)

SPINNING AND WEAVING CALCULATIONS, especially relating to Woollens. From the German of N. REISER. Thirty-four Illustrations. Tables. 160 pp. Demy 8vo. 1904. Price 10s. 6d. net. (Post free, 10s. 10d. home; 11s. abroad.)

WATERPROOFING OF FABRICS. By Dr. S. Mierzinski. Crown 8vo. 104 pp. 29 Illus. Price 5s. net. (Post free, 5s. 3d. home; 5s. 4d. abroad.)

HOW TO MAKE A WOOLLEN MILL PAY. By John Mackie. Crown 8vo. 76 pp. Price 3s. 6d. net. (Post free, 3s. 9d. home; 3s. 10d. abroad.)

YARN AND WARP SIZING IN ALL ITS BRANCHES. Translated from the German of Carl Kretschmar. Royal 8vo. 123 Illustrations. 150 pp. Price 10s. 6d. net. (Post free, 10s. 10d. home; 11s. abroad.)

(*For "Textile Soaps and Oils" see p.* 7.)

(Dyeing, Colour Printing, Matching and Dye-stuffs.)

THE COLOUR PRINTING OF CARPET YARNS. Manual for Colour Chemists and Textile Printers. By David Paterson, F.C.S. Seventeen Illustrations. 136 pp. Demy 8vo Price 7s. 6d. net. (Post free, 7s. 10d. home; 8s. abroad.)

THE SCIENCE OF COLOUR MIXING. A Manual intended for the use of Dyers, Calico Printers and Colour Chemists. By David Paterson, F.C.S. Forty-one Illustrations. **Five Coloured Plates, and Four Plates showing Eleven Dyed Specimens of Fabrics.** 132 pp. Demy 8vo. Price 7s. 6d. net. (Post free, 7s. 10d. home; 8s. abroad.)

DYERS' MATERIALS: An Introduction to the Examination, Evaluation and Application of the most important Substances used in Dyeing, Printing, Bleaching and Finishing. By Paul Heerman, Ph.D. Translated from the German by A. C. Wright, M.A. (Oxon)., B.Sc. (Lond.). Twenty-four Illustrations. Crown 8vo. 150 pp. Price 5s. net. (Post free, 5s. 4d. home; 5s. 6d. abroad.)

COLOUR MATCHING ON TEXTILES. A Manual intended for the use of Students of Colour Chemistry, Dyeing and Textile Printing. By David Paterson, F.C.S. Coloured Frontispiece. Twenty-nine Illustrations and **Fourteen Specimens of Dyed Fabrics.** Demy 8vo. 132 pp. Price 7s. 6d. net. (Post free, 7s. 10d home; 8s. abroad.)

COLOUR: A HANDBOOK OF THE THEORY OF COLOUR. By George H. Hurst, F.C.S. **With Ten Coloured Plates** and Seventy-two Illustrations. 160 pp Demy 8vo. Price 7s. 6d. net. (Post free, 7s. 10d. home; 8s. abroad.)

Reissue of

THE ART OF DYEING WOOL, SILK AND COTTON. Translated from the French of M. HELLOT, M. MACQUER and M. LE PILEUR D'APLIGNY. First Published in English in 1789. Six Plates. Demy 8vo. 446 pp. Price 5s. net. (Post free, 5s. 6d. home; 6s. abroad.)

THE CHEMISTRY OF DYE-STUFFS. By Dr. GEORG VON GEORGIEVICS. Translated from the Second German Edition. 412 pp. Demy 8vo. Price 10s. 6d. net. (Post free, 11s. home; 11s 6d. abroad.)

THE DYEING OF COTTON FABRICS: A Practical Handbook for the Dyer and Student. By FRANKLIN BEECH, Practical Colourist and Chemist. 272 pp. Forty-four Illustrations of Bleaching and Dyeing Machinery. Demy 8vo. Price 7s. 6d. net. (Post free, 7s. 10d. home; 8s. abroad.)

THE DYEING OF WOOLLEN FABRICS. By FRANKLIN BEECH, Practical Colourist and Chemist. Thirty-three Illustrations. Demy 8vo. 228 pp. Price 7s. 6d. net. (Post free, 7s. 10d. home; 8s. abroad.)

(Silk Manufacture.)

SILK THROWING AND WASTE SILK SPINNING. By HOLLINS RAYNER. Demy 8vo. 170 pp. 117 Illus. Price 5s. net. (Post free, 5s. 4d. home; 5s. 6d. abroad.)

(Bleaching and Bleaching Agents.)

A PRACTICAL TREATISE ON THE BLEACHING OF LINEN AND COTTON YARN AND FABRICS. By L. TAILFER, Chemical and Mechanical Engineer. Translated from the French by JOHN GEDDES McINTOSH. Demy 8vo. 303 pp. Twenty Illus. Price 12s. 6d. net. (Post free, 13s. home; 13s. 6d. abroad.)

MODERN BLEACHING AGENTS AND DETERGENTS. By Professor MAX BOTTLER. Translated from the German. Crown 8vo. 16 Illustrations. 160 pages. Price 5s. net. (Post free, 5s. 3d. home; 5s. 6d. abroad.)

(Cotton Spinning and Combing.)

COTTON SPINNING (First Year). By THOMAS THORNLEY, Spinning Master, Bolton Technical School. 160 pp. Eighty-four Illustrations. Crown 8vo. Second Impression. Price 3s. net. (Post free, 3s. 4d. home; 3s. 6d. abroad.)

COTTON SPINNING (Intermediate, or Second Year). By THOMAS THORNLEY. Second Impression. 180 pp. Seventy Illustrations. Crown 8vo. Price 5s. net. (Post free, 5s. 4d. home: 5s. 6d. abroad.)

COTTON SPINNING (Honours, or Third Year). By THOMAS THORNLEY. 216 pp Seventy-four Illustrations. Crown 8vo. Second Edition. Price 5s. net. (Post free, 5s. 4d. home; 5s. 6d. abroad.)

COTTON COMBING MACHINES. By THOS. THORNLEY, Spinning Master, Technical School, Bolton. Demy 8vo. 117 Illustrations. 300 pp. Price 7s. 6d. net. (Post free, 8s. home; 8s. 6d. abroad.)

COTTON WASTE: Its Production, Characteristics, Regulation, Opening, Carding, Spinning and Weaving. By THOMAS THORNLEY. Demy 8vo. About 300 pages. [*In the press.*

THE RING SPINNING FRAME: GUIDE FOR OVERLOOKERS AND STUDENTS. By N. BOOTH. Crown 8vo. 76 pages. Price 3s. net. (Post free, 3s. 3d. home; 3s. 6d. abroad.) [*Just published.*

(Flax, Hemp and Jute Spinning.)

MODERN FLAX, HEMP AND JUTE SPINNING AND TWISTING. A Practical Handbook for the use of Flax, Hemp and Jute Spinners, Thread, Twine and Rope Makers. By HERBERT R. CARTER, Mill Manager, Textile Expert and Engineer, Examiner in Flax Spinning to the City and Guilds of London Institute. Demy 8vo. 1907. With 92 Illustrations. 200 pp. Price 7s. 6d. net. (Post free, 7s. 9d. home; 8s. abroad.)

(Collieries and Mines.)

RECOVERY WORK AFTER PIT FIRES. By ROBERT LAMPRECHT, Mining Engineer and Manager. Translated from the German. Illustrated by Six large Plates, containing Seventy-six Illustrations. 175 pp. Demy 8vo. Price 10s. 6d. net. (Post free, 10s. 10d. home; 11s. abroad.)

VENTILATION IN MINES. By ROBERT WABNER, Mining Engineer. Translated from the German. Royal 8vo. Thirty Plates and Twenty-two Illustrations. 240 pp. Price 10s. 6d. net. (Post free, 11s. home; 11s 3d. abroad.)

HAULAGE AND WINDING APPLIANCES USED IN MINES. By CARL VOLK. Translated from the German. Royal 8vo. With Six Plates and 148 Illustrations. 150 pp. Price 8s. 6d. net. (Post free, 9s home; 9s. 3d. abroad.)

For contents of these books, see List III.

THE ELECTRICAL EQUIPMENT OF COLLIERIES. By W. GALLOWAY DUNCAN, Electrical and Mechanical Engineer, Member of the Institution of Mining Engineers, Head of the Government School of Engineering, Dacca, India; and DAVID PENMAN, Certificated Colliery Manager, Lecturer in Mining to Fife County Committee. Demy 8vo. 310 pp. 155 Illustrations and Diagrams. Price 10s. 6d. net. (Post free, 11s. home; 11s. 3d. abroad.)

(Dental Metallurgy.)

DENTAL METALLURGY: MANUAL FOR STUDENTS AND DENTISTS. By A. B. GRIFFITHS, Ph.D. Demy 8vo. Thirty-six Illustrations. 200 pp. Price 7s. 6d. net. (Post free, 7s 10d. home: 8s. abroad.)

(Engineering, Smoke Prevention and Metallurgy.)

THE PREVENTION OF SMOKE. Combined with the Economical Combustion of Fuel. By W. C. POPPLEWELL, M.Sc., A.M. Inst., C.E., Consulting Engineer. Forty-six Illustrations. 190 pp. Demy 8vo. Price 7s. 6d. net. (Post free, 7s. 10d. home; 8s. 3d. abroad.)

GAS AND COAL DUST FIRING. A Critical Review of the Various Appliances Patented in Germany for this purpose since 1885. By ALBERT PÜTSCH. 130 pp. Demy 8vo. Translated from the German. With 103 Illustrations Price 5s. net. (Post free, 5s. 4d. home; 5s. 6d. abroad.)

THE HARDENING AND TEMPERING OF STEEL IN THEORY AND PRACTICE. By FRIDOLIN REISER. Translated from the German of the Third Edition. Crown 8vo. 120 pp. Price 5s. net. (Post free, 5s 3d. home; 5s. 4d. abroad.)

SIDEROLOGY: THE SCIENCE OF IRON (The Constitution of Iron Alloys and Slags). Translated from German of HANNS FREIHERR V. JÜPTNER. 350 pp. Demy 8vo. Eleven Plates and Ten Illustrations. Price 10s. 6d. net. (Post free, 11s home; 11s. 6d. abroad.)

EVAPORATING, CONDENSING AND COOLING APPARATUS. Explanations, Formulæ and Tables for Use in Practice. By E. HAUSBRAND, Engineer. Translated by A. C. WRIGHT, M.A. (Oxon.), B.Sc., (Lond.). With Twenty-one Illustrations and Seventy-six Tables. 400 pp Demy 8vo. Price 10s. 6d. net. (Post free, 11s. home; 11s. 6d. abroad.)

(The "Broadway" Series of Engineering Handbooks.)

VOLUME I.—**REINFORCED CONCRETE.** By EWART S. ANDREWS, B.Sc. Eng. (Lond.). Small crown 8vo. 200 pp. 54 Illustrations and Numerous Worked Examples. Price 3s. net. (Post free, 3s. 3d. home; 3s. 6d. abroad.) [*Just published.*

VOLUME II.—**GAS AND OIL ENGINES.** [*In the press.*

VOLUME III.—**STRUCTURAL STEEL AND IRON WORK.** [*In the press.*

VOLUME IV.—**TOOTHED GEARING.** By G. T. WHITE, B.Sc. (Lond.). [*In the press.*

VOLUME V.—**STEAM TURBINES**: Their Theory and Construction. [*In the press.*

(Sanitary Plumbing, Electric Wiring, Metal Work, etc.)

EXTERNAL PLUMBING WORK. A Treatise on Lead Work for Roofs. By JOHN W. HART, R.P.C. 180 Illustrations. 272 pp. Demy 8vo. Second Edition Revised. Price 7s. 6d. net. (Post free, 7s. 10d. home; 8s. abroad.)

HINTS TO PLUMBERS ON JOINT WIPING, PIPE BENDING AND LEAD BURNING. Third Edition Revised and Corrected, By JOHN W. HART, R.P.C. 184 Illustrations. 313 pp. Demy 8vo. Price 7s. 6d. net. (Post free, 8s. home; 8s. 6d. abroad.)

SANITARY PLUMBING AND DRAINAGE. By JOHN W. HART. Demy 8vo. With 208 Illustrations. 250 pp. 1904. Price 7s. 6d. net. (Post free, 7s. 10d. home; 8s. abroad.)

ELECTRIC WIRING AND FITTING. By SYDNEY F. WALKER, R.N., M.I.E.E., M.I.Min.E., A.M.Inst.C.E., etc., etc. Crown 8vo 150 pp. With Illustrations and Tables. Price 5s. net. (Post free, 5s. 3d. home; 5s. 6d. abroad.)

THE PRINCIPLES AND PRACTICE OF DIPPING, BURNISHING, LACQUERING AND BRONZING BRASS WARE. By W. NORMAN BROWN. 48 pp. Crown 8vo. Price 3s. net. (Post free, 3s. 3d. home and abroad.) [*Just published.*

THE DEVELOPMENT OF THE INCANDESCENT ELECTRIC LAMPS. By G. BASIL BARHAM, A.M.I.E.E. Illustrated. Demy 8vo. 200 pp. Price 5s. net. (Post free, 5s. 4d. home; 5s. 6d. abroad.) [*Just published.*

For contents of these books, see List I.

WIRING CALCULATIONS FOR ELECTRIC LIGHT AND POWER INSTALLATIONS. A Practical Handbook containing Wiring Tables, Rules, and Formulæ for the Use of Architects, Engineers, Mining Engineers, and Electricians, Wiring Contractors and Wiremen, etc. By G. W. LUMMIS PATERSON. Crown 8vo. 96 pages. 35 Tables. Price 5s. net. (Post free, 5s. 3d. home ; 5s. 6d. abroad.)

[*Just published.*

A HANDBOOK ON JAPANNING AND ENAMELLING FOR CYCLES, BEDSTEADS, TINWARE, ETC. By WILLIAM NORMAN BROWN. 52 pp. and Illustrations. Crown 8vo. Price 2s. net. (Post free, 2s. 3d. home and abroad.)

THE PRINCIPLES OF HOT WATER SUPPLY. By JOHN W. HART, R.P.C. With 129 Illustrations. 177 pp. Demy 8vo. Price 7s. 6d. net. (Post free, 7s. 10d. home ; 8s. abroad.)

(Brewing and Botanical.)

HOPS IN THEIR BOTANICAL, AGRICULTURAL AND TECHNICAL ASPECT, AND AS AN ARTICLE OF COMMERCE. By EMMANUEL GROSS, Professor at the Higher Agricultural College, Tetschen-Liebwerd. Translated from the German. Seventy-eight Illustrations. 340 pp. Demy 8vo. Price 10s. 6d. net. (Post free, 11s. home ; 11s. 6d. abroad.)

INSECTICIDES, FUNGICIDES AND WEED-KILLERS. Demy 8vo. About 500 pp.

[*In the press.*

(Wood Products, Timber and Wood Waste.)

WOOD PRODUCTS: DISTILLATES AND EXTRACTS. By P. DUMESNY, Chemical Engineer, Expert before the Lyons Commercial Tribunal, Member of the International Association of Leather Chemists; and J. NOYER. Translated from the French by DONALD GRANT. Royal 8vo. 320 pp. 103 Illustrations and Numerous Tables. Price 10s. 6d. net. (Post free, 11s. home ; 11s. 6d. abroad.)

TIMBER: A Comprehensive Study of Wood in all its Aspects (Commercial and Botanical), showing the different Applications and Uses of Timber in Various Trades, etc. Translated from the French of PAUL CHARPENTIER. Royal 8vo. 437 pp. 178 Illustrations. Price 12s. 6d. net. (Post free, 13s. home; 14s. abroad.)

THE UTILISATION OF WOOD WASTE. Translated from the German of ERNST HUBBARD. Crown 8vo. 192 pp. Fifty Illustrations. Price 5s. net. (Post free, 5s 4d. home; 5s. 6d. abroad.)

(*See also Utilisation of Waste Products, p.* 9.)

(Building and Architecture.)

ORNAMENTAL CEMENT WORK. By OLIVER WHEATLEY. Demy 8vo. 83 Illustrations. 128 pp. Price 5s. net. (Post free, 5s. 4d. home; 5s. 6d. abroad.) [*Just published.*

THE PREVENTION OF DAMPNESS IN BUILDINGS; with Remarks on the Causes, Nature and Effects of Saline, Efflorescences and Dry-rot, for Architects, Builders, Overseers, Plasterers, Painters and House Owners. By ADOLF WILHELM KEIM. Translated from the German of the second revised Edition by M. J. SALTER, F.I.C., F.C.S. Eight Coloured Plates and Thirteen Illustrations. Crown 8vo. 115 pp. Price 5s. net. (Post free, 5s. 3d. home; 5s. 4d. abroad.)

HANDBOOK OF TECHNICAL TERMS USED IN ARCHITECTURE AND BUILDING, AND THEIR ALLIED TRADES AND SUBJECTS. By AUGUSTINE C. PASSMORE. Demy 8vo. 380 pp. Price 7s. 6d. net. (Post free, 8s. home; 8s. 6d. abroad.)

(Foods, Drugs and Sweetmeats.)

FOOD AND DRUGS By E. J. PARRY, B.Sc., F.I.C., F.C.S.

Volume I. The Analysis of Food and Drugs (Chemical and Microscopical). Royal 8vo. 724 pp. Price 21s. net. (Post free, 21s. 8d. home; 22s. abroad.)

Volume II. The Sale of Food and Drugs Acts, 1875-1907. Royal 8vo. 184 pp. Price 7s. 6d. net. (Post free, 7s. 10d. home; 8s. abroad.) [*Just published.*

THE MANUFACTURE OF PRESERVED FOODS AND SWEETMEATS. By A. HAUSNER. With Twenty-eight Illustrations. Translated from the German of the third enlarged Edition. Crown 8vo. 225 pp. Price 7s. 6d. net. (Post free, 7s. 9d. home; 7s. 10d. abroad.)

RECIPES FOR THE PRESERVING OF FRUIT, VEGETABLES AND MEAT. By E. WAGNER. Translated from the German. Crown 8vo. 125 pp. With 14 Illustrations. Price 5s. net. (Post free, 5s. 3d. home; 5s. 4d. abroad.)

For contents of these books, see List III.

(Dyeing Fancy Goods.)

THE ART OF DYEING AND STAINING MARBLE, ARTIFICIAL STONE, BONE, HORN, IVORY AND WOOD, AND OF IMITATING ALL SORTS OF WOOD. A Practical Handbook for the Use of Joiners, Turners, Manufacturers of Fancy Goods, Stick and Umbrella Makers, Comb Makers, etc. Translated from the German of D. H. SOXHLET, Technical Chemist. Crown 8vo. 168 pp. Price 5s. net. (Post free, 5s. 3d. home; 5s. 4d. abroad.)

(Celluloid.)

CELLULOID: Its Raw Material, Manufacture, Properties and Uses. A Handbook for Manufacturers of Celluloid and Celluloid Articles, and all Industries using Celluloid: also for Dentists and Teeth Specialists. By Dr. Fr. BÖCKMANN, Technical Chemist. Translated from the Third Revised German Edition. Crown 8vo. 120 pp. With 49 Illustrations. Price 5s. net. (Post free, 5s. 3d. home; 5s. 4d. abroad.)

(Lithography, Printing and Engraving.)

PRACTICAL LITHOGRAPHY. By ALFRED SEYMOUR. Demy 8vo. With Frontispiece and 33 Illus. 120 pp. Price 5s. net. (Post free, 5s. 4d. home; 5s. 6d. abroad.)

PRINTERS' AND STATIONERS' READY RECKONER AND COMPENDIUM. Compiled by VICTOR GRAHAM. Crown 8vo. 112 pp. 1904. Price 3s. 6d. net. (Post free, 3s. 9d. home; 3s. 10d. abroad.)

ENGRAVING FOR ILLUSTRATION. HISTORICAL AND PRACTICAL NOTES. By J. KIRKBRIDE. 72 pp. Two Plates and 6 Illustrations Crown 8vo. Price 2s. 6d. net. (Post free, 2s. 9d. home; 2s. 10d. abroad.)

(*For Printing Inks, see p.* 4.)

(Bookbinding.)

PRACTICAL BOOKBINDING. By PAUL ADAM. Translated from the German. Crown 8vo. 180 pp. 127 Illustrations. Price 5s. net. (Post free, 5s. 4d. home; 5s. 6d. abroad.)

(Sugar Refining.)

THE TECHNOLOGY OF SUGAR: Practical Treatise on the Modern Methods of Manufacture of Sugar from the Sugar Cane and Sugar Beet. By JOHN GEDDES MCINTOSH. Second Revised and Enlarged Edition. Demy 8vo. Fully Illustrated. 436 pp. Seventy-six Tables. 1906. Price 10s. 6d. net. (Post free, 11s. home; 11s. 6d. abroad.)

(*See "Evaporating, Condensing, etc., Apparatus," p.* 9.

Spotted something that isn't quite right?

Let us know and we'll fix it for future reprints.
And as a thank you for your diligence, we'll
send you a free Sothis book of your choice.

publisher@sothis.press

This page intentionally left blank

Made in United States
Orlando, FL
10 November 2026

90505159R00129